WINDING DOWN

THE PRICE OF DEFENSE

WINDING DOWN

THE PRICE OF DEFENSE

The Boston Study Group

W. H. FREEMAN AND COMPANY
SAN FRANCISCO

This book was previously published by Times Books, a division of Quadrangle/The New York Times Book Company, Inc.

Library of Congress Cataloging in Publication Data

Boston Study Group.
 Winding down.

   Reprint. Originally published: The price of defense.
New York : Times Books, c1979.
   Bibliography: p.
   Includes index.
   1. United States—Military policy.   2. Soviet Union
—Military policy.   3. United States—Armed Forces—
Appropriations and expenditures.   4. Soviet Union—
Armed Forces—Appropriations and expenditures.
5. Disarmament.   I. Title.
UA23.B725   1982   355'.033073   82-18214
ISBN 0-7167-1498-1 (pbk.)

Printed in the United States of America

1 2 3 4 5 6 7 8 9 0   MP   0 8 9 8 7 6 5 4 3 2

For the Children

CONTENTS

FOREWORD

The U.S. government is presently committed to a buildup of its military expenditures steeper than any that has marked the arms race of the last four decades. The cumulative military outlay over the four years of the Reagan Administration will exceed $1,000 billion. Every service is in line for new weapons. They will deploy the latest advances in the continuing revolution in weapons technology that now takes its impetus from control technology with integrated solid-state electronics. The emblematic innovation of this round in the cycle will be the long-range cruise missile. Its advent is as portentous in its turn as that of the jet aircraft and the intercontinental ballistic missile.

The projected increase in the destructive power of the U.S. armed forces is paced by the proliferation of nuclear warheads into every mode and size of delivery vehicle. Most visible, of course, are those installed in the strategic weapons systems. The triad of bombers, land-based intercontinental ballistic missiles and submarine-launched intercontinental ballistic missiles is now diversified by air- and submarine-launched cruise missiles. The number of free-fall nuclear bombs and nuclear warheads addressed to targets in the U.S.S.R. from outside the continent of Europe is scheduled to increase from 9,000-plus to 11,000-plus by 1985 even under the terms of the unratified and precarious Salt II treaty. Without the treaty it could go to 17,000-plus. To these 1985 totals should be added about 1,000 more nuclear warheads to be delivered to strategic targets by intermediate-range Pershing ballistic missiles and first-generation cruise missiles deployed in Europe. Ignored by all the arms-control negotiations undertaken so far are some thousands of nuclear explosive charges to be manufactured for installation in tactical battlefield weapons, in free-fall bombs, howitzer shells and in missiles of less than strategic range.

By the testimony of those in charge, in and out of uniform, and from the accidental or deliberate leakage of Pentagon papers, there is evidence of a metamorphosis in U.S. military doctrine governing the prospect and conduct of thermonuclear war. The doctrine of mutual assured destruction that has rationalized prior increases in the country's nuclear armament is now transcended. In its place the feasibility of a counterforce first-strike, aimed by one or the other party at its enemy's nuclear arsenal, is argued. From there concern is directed to preparation for fighting a "protracted nuclear conflict." It may be recalled, however, that the exchange of a mere 1,000 high-yield missiles in a breakdown of the old mutual-assured-destruction standoff had once been reckoned as prospectively the greatest calamity in the history of life. The *Defense Guidance: Five-Year Plan* assures the readers for whom it was first intended that a fighting fraction of the nation's nuclear arsenal will survive a first-strike with no more than an "acceptable" twenty million civilian casualties. Beyond that horizon, the Secretary of Defense acknowledges his worry about "the end of the world."

In principle, the commitment of the government of the United States, arrived at by the democratic process, declares the commitment of the people. Each citizen, whether committed consciously or not, has personal complicity in the policies and actions of the government—and, with respect to present military policy, a direct exposure to the prospective physical consequences. Where does a concerned citizen turn for understanding of the highly technical and repugnant matters that have lead presumably rational and responsible fellow Americans in public office to the costly and desperate course of their Five-Year Plan?

The primary source of objective information for public deliberation on military questions is, of course, the military establishment. To suggest that there ought to be another source casts no shadow on the professionalism, expertise or probity of these public servants. From everyday experience with other bureaucracies, public and private, that manage the affairs of industrial civilization, citizens have come to know that these organizations develop internal imperatives often stultifying of their performance. The organization of the single Department of Defense thirty-seven years ago has not suppressed the rivalries among the uniformed services that amplify the drive for bureaucratic aggrandizement resident in each of them. Outside the government, the industries that supply the weapons constitute a powerful economic and political interest in the growth of the military establishment.

One asks a second opinion on military policy as on the recommendation of a surgeon. A surgeon offers surgery as his remedy; the soldier, force and violence. In acquitting his awful responsibility the soldier must make the worst-case analysis: the enemy's weapons and strategy will work flawlessly; ours must be discounted for mechanical failure and human frailty. His claim on a nation's human and material resources is absolute and without limit. That is why the military is placed by the U.S. Constitution under civilian command.

What is more, the preoccupation of the Pentagon with counterforce strategy raises the suspicion that the generals and admirals are once again fighting the last war. Those who remember Pearl Harbor will recall that the U.S. armed forces were caught with their ships at anchor and their planes on the ground all the way from the Philippines to Hawaii on December 7, 1941. Hiroshima and Nagasaki give that memory a lurid new scenario. There is grim logic to the military preoccupation with security of the arsenal; for the military cannot protect the population. The counterforce arms race threatens, however, to escalate the possible beyond the probable to the inevitable.

Fortunately, there are alternative sources of objective information and independent counsel on military questions. The ongoing technological revolution in warfare has engaged a great many trained minds from outside the military establishment. The nuclear weapons and the propulsion and control technologies that deliver them come from the engineering frontiers of physics. University professors must be counted among the nation's principal armorers. As science advisers to the military services, the intelligence agencies, the White House and the Congress, they have perforce been engaged in the study of war and the uses of the weapons that have issued from their work. They are fully qualified for service as military advisers to their fellow citizens.

Scientific counsel and testimony are usually rendered *ad hoc* with respect to specific decisions: then, whether to proceed with the development of the hydrogen bomb; now, the basing of the MX Missile. That counsel has not always been heeded by the decisionmakers in the government. For that reason alone, such arguments ought to be heard more widely by their fellow civilians. In this book, the Boston Study Group have undertaken a more comprehensive task. They have assembled a structured picture of the country's entire military establishment. It shows a system driven by technological change that tends to outrun policy and the accomodation of vested interest. As the intercontinental ballistic missile with its nuclear warheads made the offense paramount in strategic (read: population) warfare, so this book shows control technology steadily making the defense paramount in battle. The sinking of an Argentine cruiser and a British destroyer by "smart" missiles in the Falkland Islands war has put in doubt again the wisdom of heavy investment in aircraft carriers and tanks.

This picture of the military establishment displays its physiology as well as its anatomy. It explains the system in terms of the missions to which it is committed. Such analysis exposes the fatuity of the sheer numbers—the "bean counting" of men, tanks, ships, missiles, throw-weight—by which the progress of the arms race is commonly reckoned and a costly shopping list sold to the people and their representatives from year to year.

The missions for which a military establishment is created and maintained express the political objectives of the nation. The self-governing U.S. citizen, with or without a bent for technical matters, has the duty to consider the vision of the role of our nation in world politics that is incarnated in our armed forces. If "defense" were truly the mission, this book sets out a design for a cheaper and much safer U.S. establishment for our defense by military means.

Gerard Piel, Publisher
Scientific American
September 1982

PREFACE TO THE 1982 EDITION

This book presents a critical alternative analysis of the U.S. military budget, now projected to approach $400 billion in the year 1987. *Winding Down* argues that U.S. security can be improved by *reducing* defense spending over several years by about 40 percent: that doing so will provide us with a better, more effective, safer, and cheaper national defense. Our aim in working out the entire American military budget in some detail, using FY 1978 as an example, has been to equip the general reader with the information needed for a grasp of these issues, which affect us all.

Since 1979 American national security has fully entered the new phase of post-Vietnam war planning: currently one of annual double-digit real increases in military spending, of new generations of sophisticated nuclear technology, of expansive nonnuclear force deployments including large interventionary capabilities. Present plans are reported to call for many thousands of new nuclear warheads, half-a-dozen major new nuclear weapon delivery systems, new Army and Marine divisions, new air wings, and a 650-ship naval fleet including 22 aircraft carriers by 1987. The U.S. forces do not yet reflect these spending increases. Most of the increase, apart from pay, is an investment for the military future. But missiles, submarines, planes, and the rest take time to build. The major components of our forces in early 1982 are surprisingly like those of 1978; but today the guns have more shells to fire, the divisions more and newer equipment. The fighter squadrons have newer planes and spend more hours aloft, and there are new ships and attack submarines. It seems likely that by the late 1980s there would have to be a substantial increase in uniformed personnel as well.

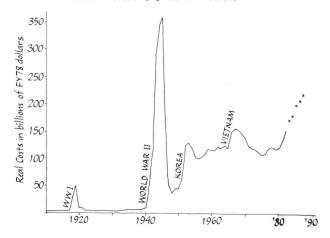

US MILITARY COSTS OVER THE YEARS

The real military expenditures of the United States—past, present, and proposed—with inflation removed are shown in this figure, an update of Figure 1 in this book. The dots are the expenditures planned by the Reagan Administration; the 1983 numbers are still unsettled as we write in June of 1982. Spending is turning sharply upward; the implied growth averages about 8 percent yearly for the next five years. The correction for inflation allows a rough comparison across the years, but it is to be recalled that the United States is today a much bigger country, both in population and in its economic base, than it was during World War II.

A comparable measure can be offered for the military spending of the U.S.S.R. That curve shows a steady growth since the early 1960s, more nearly a fixed annual increment than a compound percentage increase, which amounts to about 4 percent per year now in the early 1980s. (This estimate, equalling one-half the U.S. military growth rate, is the official one made by the Department of Defense and the CIA; but see Chapter 4 for our critique of their methods.)

The current composition and some of the future plans for our forces are to be found each year in the *Annual Report of the Secretary of Defense to Congress*, the most compact and authoritative source for readers wishing to update in detail the information in this book.

The Boston Study Group
Cambridge, Massachusetts
July 1982

PREFACE TO THE 1979 EDITION

The Boston Study Group is a private and informal group concerned with the risks, the costs, the adequacy, and the excesses of United States military policy today.

We came together with a few other interested individuals at the beginning of 1975 to discuss the forces and spending which would safely support a truly defensive U.S. military posture. Among us are both experienced military analysts and citizens looking for the first time at these problems. We brought a variety of viewpoints and experiences, with service amidst old wars and newer tensions, familiarity with Capitol lobbies, and with the corridors of the Pentagon. We read, reported, analyzed, discussed, drafted, and redrafted until reasoned consensus could be found.

Our sources, which we cite, are all public and unclassified. The most important of these are the official reports of the U.S. Department of Defense and its agencies and the reports of Congressional hearings and staff. The international scholarly literature and the trade journals are valuable as well, and we have consulted a good many experts on various special questions. The results of our work are set out in this book, meant for the attentive general reader. The reader will find that along with our arguments we provide sufficient background material so that the book can also serve as a primer of modern warfare.

We were not always alone in our deliberations, but have often benefited from the comments of others. We are grateful to George Leppert and Will Watson, who first brought us together and who participated for a long time. Others who deserve much thanks for their contributions, but who are not responsible for the views expressed in the book, are Tom Dine, George Rathjens, and particularly Ronald Siegel. We are also grateful to many friends and colleagues for useful comments on the manuscript. Robert Schaeffer provided editorial assistance, and Fred Kaplan provided research assistance. Margaret Harrigan and Barbara Sleeth typed and compiled the final text. A grant from the Arms Control Project of the M.I.T. Center for International Studies to cover typing and editorial assistance is gratefully acknowledged.

The Boston Study Group
Cambridge, Massachusetts
January 1978

A NOTE ON SECRECY

Many people believe that official secrecy prevents any reliable outside study of the American military through public sources alone. No one who tries conscientiously will agree with that view. A great deal of information is freely published in such detail and in so many corroborative ways that little doubt remains as to its accuracy. Of course there are material gaps in published information: though in detail significant, they are in no way major.

The true barrier to reliable criticism of U.S. military policy is not secrecy but the sheer complexity of American military organizations and their purposes. A rational analysis must consider the world of international affairs, modern technology, and the perceptions of planners in other countries. Even harder, it must project these into the future. Despite the difficulties, we have felt the need to be specific in our critique of U.S. military forces. We did not undertake to make a generalized review, but to give quantitative detail adequate to make plain our view of the present forces and our suggestions for improvements in them. If here or there we ask for a change already planned, or mistakenly estimate details of some particular new weapon or military decision, we will be glad to have notice of such errors. Such details are in any case not central to our argument. For what we offer is not meant as an exact description, but a reasonably accurate picture of the present forces and of a future military structure of a kind to which we could better entrust our national safety.

THE AUTHORS

Randall Forsberg worked at the Stockholm International Peace Research Institute (SIPRI) from the inception of its arms control research program in 1968 until her return to the United States in 1974 to study at the Massachusetts Institute of Technology. She still contributes to the *SIPRI Yearbook of World Armaments and Disarmament*, where she has written on U.S. and Soviet military research and development, the global spread of defense industries, and strategic nuclear weapons. She is the founder and Director of the Institute for Defense and Disarmament Studies in Brookline, Massachusetts, and she helped start the Nuclear Weapons Freeze Campaign.

Martin Moore-Ede, Professor of Physiology at the Harvard Medical School, has a long-term interest in the social implications of science and science policy.

Philip Morrison, Institute Professor of Physics at the Massachusetts Institute of Technology, book review editor of *Scientific American*, and past Chairman of the Federation of American Scientists, has written on nuclear war and on aspects of U.S. military policy ever since his days with the Manhattan Project, 1942–1946, where he took a responsible part in the first test and the first use of nuclear weapons.

Phylis Morrison is a teacher, writer, and illustrator.

George Sommaripa is a politician trained in law and history. He was a candidate for the U.S. Senate from Massachusetts in 1972 and is a member of the Council for a Nuclear Weapons Freeze, Boston.

Paul F. Walker holds a Ph.D. in political science from the Massachusetts Institute of Technology and is Director of Education and Research for the Physicians for Social Responsibility. He is a former intelligence specialist in Soviet affairs with the U.S. Army Security Agency and has worked for the U.S. Arms Control and Disarmament Agency. He is author of "Precision-Guided Weapons" (*Scientific American*, August 1981) and a regular contributor to the annual survey of *World Military and Social Expenditures*.

THE PANOPLY OF THE UNITED STATES

STRATEGIC FORCES To deter & defend against long-range nuclear attack.	
Strategic offensive forces To deter foreign nuclear attacks on the USA by threatening retaliation in kind.	Submarine-launched ballistic weapons-SLBMs Land-based inter-continental ballistic weapons- ICBMs. Long-range bomber aircraft.
Strategic defensive forces To warn of any foreign nuclear attack & to defend against bombers.	Satellites & radars. Interceptor aircraft.

GENERAL PURPOSE FORCES To protect US allies, preserve the freedom of the seas & permit military intervention abroad.	

Land combat forces
  Ground troops
    Mainly against the USSR in Europe.
	Armored & Mechanized divisions & other units.

Mainly against weaker powers elsewhere.	Infantry & Airborne divisions & other units.
Transport Intercontinental airlift.	Jets, for troops & equipment, including tanks.
Airlift within any theater of combat.	Medium range aircraft for troops & equipment.
Sealift	Cargo ships to carry fuel, supplies & equipment.

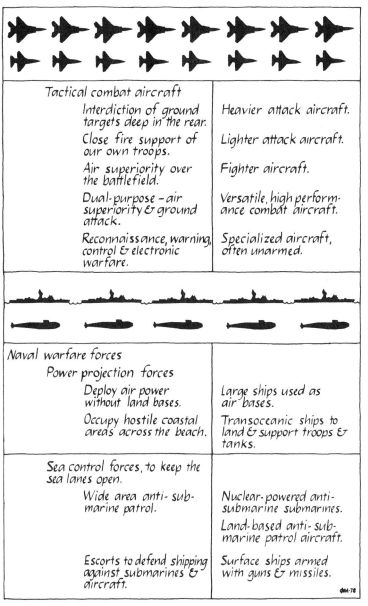

Tactical combat aircraft

Interdiction of ground targets deep in the rear.	Heavier attack aircraft.
Close fire support of our own troops.	Lighter attack aircraft.
Air superiority over the battlefield.	Fighter aircraft.
Dual-purpose – air superiority & ground attack.	Versatile, high performance combat aircraft.
Reconnaissance, warning, control & electronic warfare.	Specialized aircraft, often unarmed.

Naval warfare forces

Power projection forces

Deploy air power without land bases.	Large ships used as air bases.
Occupy hostile coastal areas across the beach.	Transoceanic ships to land & support troops & tanks.

Sea control forces, to keep the sea lanes open.

Wide area anti-submarine patrol.	Nuclear-powered anti-submarine submarines.
	Land-based anti-submarine patrol aircraft.
Escorts to defend shipping against submarines & aircraft.	Surface ships armed with guns & missiles.

The chart outlines the functions and nature of the combat forces of the United States. Figures throughout the book offer numerical details of these forces. STRATEGIC FORCES: Figure 15. LAND COMBAT FORCES: ground troops, Figures 27 and 49. Transport, Figure 55. Aircraft, Figures 29 and 33. NAVAL WARFARE FORCES: Ships and aircraft, Figure 33.

WINDING DOWN

THE PRICE OF DEFENSE

PART I
OVERVIEW:
TOWARD A SAFER
MILITARY POLICY

The full titles for sources listed in abbreviated form below the figure captions, as well as the sources and notes for the text (referenced by page), follow the text, beginning on page 315.

1

AN INDEPENDENT VIE ⟍
OF U.S. DEFENSE NEEDS

THE NEED FOR ANALYSIS

This book provides a concise yet comprehensive analysis of U.S. military needs. It describes the forces and spending needed to defend the United States, to support the defense of Western Europe, Japan, and Israel, and to deter war.

A large part of the present military establishment is not needed for these purposes. It is left over from the cold war military stance adopted by the United States in the late 1940s. The cold war justified a large standing army, unprecedented in this country in peacetime. It required military budgets which, like ocean swells at high tide, rose and fell around a far higher mark than the pre-World War II level (Figure 1). It produced a new range of military activities, which reached deep into the industrial, intellectual, and political fabric of the nation. Major events of the 1960s—the Sino-Soviet split, détente, and American reaction against involvement in Vietnam—should have spurred a fundamental rethinking of the premises of this policy. Yet such thought and changes as have occurred inside and outside the Pentagon in the aftermath of Vietnam have been superficial and minor.

Since 1972, the U.S. military—and, for that matter, the Soviet military—have been "settling down for the long haul." In establishing the paths of their future development, they have *not* moved in new directions appropriate to current and likely future global relations; nor has there been a serious effort to reach a stable peace which is preserved with much lower levels of arms. Instead, the military establishment is in the process of becoming entrenched. It is perpetuating force and spending levels, and rates of technical advance in weaponry, which were fixed at an earlier time, under different circumstances and for reasons which no longer hold.

3

US MILITARY COSTS OVER THE YEARS

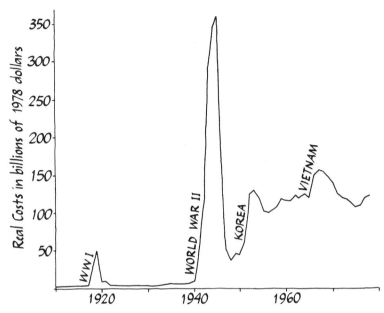

FIG. 1

The economic costs of war are not the most significant costs, but they offer one clear measure. The graph shows each year's U.S. government expenditure for military purposes, adjusted for inflation. The costs are expressed in terms of today's buying power of the dollar, using standard adjustments of the U.S. Department of Commerce.

The peaks of the four hot wars stand out plainly. The central concern is the plateau on which we now stand, unprecedented in history. In effect, the Korean War peak never came down; in peacetime real costs remain near that peak to this day.

From HISTORICAL STATISTICS OF THE US: up to 1970.
Defense outlays, series Y 458 to 460.
Deflator (implicit price index), series F5.
Data and deflator since 1970 fitted to DEFENSE SECRETARY FY 1978, p. C-1.

The perpetuation of the status quo is not surprising. Most of the experts who study and plan military forces are persons disposed to favor large military budgets. Among the public, even those well informed, there persists an exaggerated image of the military threat posed by the Soviet Union. In Washington, members of Congress find that the seven million constituents on the defense payroll form a powerful and alert "defense producers' lobby" for high military spending. The average citizen's interest in a minimum military budget and a safe, reasonable military policy are not reflected in an equally effective "defense consumers' lobby," aimed at keeping a close check on the military.

Yet it is vital to reduce the forces and resources of the military. The possibility of large-scale use of nuclear weapons poses a daily threat many times worse than any danger mankind has ever encountered. There is no defense against these weapons, nothing that can stop them once the political decision to use them has been made. The only route to safety in this realm is through reductions in the number of weapons.

Military forces as large and diversified as those possessed by the U.S.A. are not needed simply for defense against aggression. Their objective is to permit military operations in all types of conflict and in all parts of the globe, thereby providing a political advantage in the conduct of international relations and in international crises. The plethora of military forces which stand ready to back up U.S. diplomacy constitute a sort of tinderbox. But a tinderbox style of diplomacy is risky. Its twin goals—preserving the peace and coming out ahead in international tests of will—are, in fact, in conflict. For this and other reasons, such a policy may fail to keep the peace. Historically, high levels of arms are associated with the outbreak of war, not the preservation of peace. If peacetime levels of arms are high, the damage done in the event of war is likely to be greater, especially in this age of rapid destruction. Moreover, maintaining forces not required for defense invites more frequent military intervention. It does this directly by providing the tools for such action, and it does so indirectly as well. The magnitude of the military establishment, with its great claim on the scientific, intellectual, and professional resources of the nation, tends to alter the attitude of politicians, academics, and bureaucrats toward the acceptability of using force for political ends. When such a shift occurs, the distinction between offense and defense is lost in increasingly diffuse and all-encompassing objectives, which are argued to be required for national security and, in that sense, defensive.

Despite the dangerous and costly aspects of U.S. military policy, the overarching rationales for the forces have rarely been discussed since the withdrawal from Vietnam. What has been lacking is a coherent alternative to the unchanging stance of the Pentagon. Public concern with military matters has been confined to individual weapons or foreign bases or bizarre instances of waste. Little attention has been given to analysis on the comprehensive level, where the most telling policy and budget decisions are made. The broad links from major military forces to policy goals, on the one hand, and to alternative levels of military spending, on the other hand, have not been made clear.

This book is intended to fill that gap in public information. It delineates the forces needed for acceptable foreign policy goals in the world of the late 1970s and 1980s. The analysis is performed in four steps. First, the military establishment is divided into its main components, which are relevant to different policy goals. Next, the political and military justifications for each component are reviewed. Changes are then proposed wherever we judge that the forces are out of line with the goals or the goals themselves are unacceptable. Finally, the costs of the components we propose to retain are added up, giving a total alternative defense budget.

OUR RECOMMENDATIONS

Briefly put, we recommend that the half of the present military establishment aimed mainly at responding to aggression by the Soviet Union should be left unchanged or improved, while most of the other half, not useful for countering Soviet military threats or accomplishing other acceptable goals, should be eliminated.

The three major items which we cut back are:
- *the vast excess in the quantity of nuclear weapons*, over and above the number needed to deter a nuclear attack;
- most of the *aircraft carriers, amphibious landing ships,* and *lightly equipped land-combat forces*, which are primarily useful not against the U.S.S.R. but against the lesser military powers in the poorer half of the world, like Vietnam; and
- the *unnecessary investment in development of new weaponry*, which has long made the United States the driving force behind the rapid, destabilizing, and costly advances in world military technology.

The forces we propose to retain are:
- a relatively *small but invulnerable nuclear-weapon force* to

deter Soviet nuclear attacks by threatening retaliation;
- the *heavily equipped land-combat forces* presently assigned to help defend Western Europe against possible Soviet aggression together with
- *most of the current tactical combat aircraft*, which are intended to provide air cover in the event of a war in Europe and to protect the ocean approaches to Japan; and
- *a largely unchanged force of surface ships and attack submarines*, to protect the freedom of the seas.

The changes we propose are meant to be introduced over a period of 5 to 10 years. This gradual transition is intended to preserve international stability and to permit a smooth conversion to civilian employment. We also assume that a major revamping of U.S. military policy would be preceded by several years of national publicity, reflection, and debate. Thus, data from the 1978 budget and force structure are used only as an illustrative point of departure. The idea is to make clear the sort of changes that should be made in 5 to 10 years.

LONGER-TERM GOALS

Looking ahead, the military stance we recommend is not intended to be a permanent policy or an ultimate goal. Like the process of arms competition, the process of demilitarization is a dynamic one, involving interaction with other countries. Our proposals are based on projections of the military capabilities of other nations, especially the Soviet Union. They are subject to change should foreign developments warrant. A significant *increase* in Soviet conventional forces confronting U.S. allies, for example, might require a change in our program. In that case, the change would *not* be to restore the force components we eliminate. Instead, we would strengthen the components we retain, which are more suitable for countering Soviet moves. *Reductions* in Soviet forces, on the other hand, might permit further decreases on the U.S. side.

Since the rationales for both U.S. and Soviet forces involve other countries, an even better response to Soviet cuts might be reductions in the forces of U.S. allies, particularly in Europe. After a long period of rearmament, West Germany now has the world's third highest military budget and third most powerfully equipped land army. France is fourth in both respects. The plain from the Urals to the Atlantic is the most heavily armed region of the globe.

Moreover, centuries of warfare across France, Germany, Poland, and Russia have left a quiescent legacy of hostility among these nations. This hostility feeds on the heavy militarization of the area, even while it provides most of the justification for the German and French forces and for a large part of Soviet and American arms.

In theory, the United States might disengage from this old arena of conflict and confine its military forces to those needed for the continental defense of this country. However, U.S. disarmament of that magnitude, undertaken at one fell swoop, would probably result in compensating increases in the armaments of the other Western industrial giants—West Germany and Japan. This would intensify the German–Soviet standoff and probably worsen Soviet–Japanese relations, thereby increasing the likelihood of war. The alternative which we prefer is for the United States to take a more patient and farsighted course, working for an overall reduction in militarization rather than simply shifting the burden to other countries.

With this view, we recommend, as a first step, reductions in the surplus, the dangerous, and the interventionist elements of U.S. forces. We do not propose to reduce the components intended to help defend allies from external aggresion. Thus, our proposal by no means represents the ultimate cutback of U.S. military forces to the minimal corps needed for a purely national defense.

Speculating about the future, we have some hope that a major U.S. cutback would not be merely a one-shot reduction, to be eroded by some inexorable global increase in armaments. Since World War II, the United States, more than any other country, has led the world toward increased militarization. It has continually advanced the frontier of military technology. It has aided the expansion of foreign military forces and foreign defense industries. Most important, it has constantly upgraded its own military forces to maintain the image and reality of military superiority over the U.S.S.R. (an edge which the Soviet Union in turn constantly tries to narrow), rather than allow the U.S.S.R. to achieve something widely perceived as an even balance of forces. The United States has permitted its effort to keep the advantage to displace trends toward a more stable situation from which mutual force reductions might be worked out.

Relying too heavily and indiscriminately on the usefulness of military force in the conduct of foreign policy, the United States has set a standard which encourages militarization in other countries. In doing so, it has not only ignored the difficult consequences of widespread increases in nonnuclear arms, but also invited the profound dangers of the proliferation of nuclear weapons to many other countries.

If we are to leave our children a safer and freer world, rather than a more dangerous and restrictive one, we must change the steady trend in world armaments. As the leader in the world arms race, and pacesetter for the military and foreign policies of the Western world, the United States has both the responsibility and the opportunity to take the initiative in attempting to shift the fundamental course of global military policies.

INTERNATIONAL NEGOTIATIONS AND RESPONSES

We do not suggest that the United States should cease to negotiate arms reductions with the Soviet Union and coordinate proposed changes in U.S. forces with the plans of allies. We do believe, however, that more attention must be given to what the United States might safely accomplish on its own, and less reliance placed on negotiations as the sole route to arms reductions, than has been the case hitherto.

During the last decade, international negotiations have addressed reductions in those military forces which account for the bulk of world armaments and which include the most dangerous arms. These negotiations are, first, the bilateral talks to limit and reduce U.S. and Soviet strategic nuclear weapons (SALT, the Strategic Arms Limitations Talks); and, second, the multilateral talks to limit and reduce the conventional and tactical nuclear forces in Europe, including not only the forces of Eastern and Western European countries, but also the reinforcing forces of the United States and the Soviet Union (MFR, the Mutual Force Reduction talks).

In both negotiations, political and bureaucratic in-fighting within the countries involved has resulted in nonproductive negotiating proposals. The proposals of each side have tended to enhance their own advantage at the expense of the other side, and they have generally been rejected by the other side precisely for this reason. In both cases, furthermore, increases in the very forces under negotiation have been made by the two sides during the course of the talks, so as to provide better bargaining chips. In neither case have any meaningful force reductions been achieved. Indeed, in the MFR talks, there is little evidence that any of the countries involved is yet seriously interested in substantial force reductions. This is not surprising, since there is little public awareness of the negotiation or public pressure for arms reduction to be accomplished through this mechanism, despite the fact that fear of another major war in Europe is the primary rationale for at least half of U.S. and Soviet military forces and spending.

The sad record of the SALT and MFR negotiations has prompted us to look for initiatives that the United States can undertake on its own. As already indicated, the U.S. force reductions that we propose are carefully evaluated, not only with regard to the extent to which they place a constraint on U.S. military operations, but also with respect to their likely impact on the Soviet Union and their implications for U.S. allies. We have taken into account the foreign military responses that we believe would be most likely. We have also considered the possibility of less favorable foreign responses—in particular the possibility of increased militarization on the part of allies or the Soviet Union or both.

The chance of increased nuclear proliferation is particularly troublesome. It is possible that one or another of the countries bordering the Soviet Union or China—such as West Germany, Japan, South Korea, or Taiwan—will acquire nuclear weapons if the United States reduces or withdraws military forces which, in their view, support their defense. In the cases of West Germany and Japan, we have gone to some length to avoid providing any incentive or excuse for increased armament or acquisition of nuclear weapons. In other cases, the possibility of proliferation has weighed less heavily in our estimates and proposals, for two reasons. First, to the extent that other countries' nuclear decisions are significantly influenced by U.S. policy (which may be minimally or not at all), we believe that the current U.S. policy may provide as much incentive to "go nuclear" as an attempt to reverse that policy. Second, when combined with the reduced danger of war and the enhanced prospects for arms control among the industrial nations, the risks our proposals may raise in regard to third world nuclear proliferation are, on balance, less than those of continuing the present policy.

At the least, even if the U.S. initiative fails to spark a constructive response abroad, we will have removed from this country the onus of contributing to the forward momentum of militarization. At most, if we succeed in a reversal, an eventual change from the 5 to 10 percent of world income spent on arms since World War II to the 3 percent which characterized the period between the world wars—or even down to the 1 percent which prevailed before World War I—is entirely possible. This would transform the world in innumerable indirect ways. Most important, it would, in our view, reduce the likelihood of what could be a final war among the industrial nations. It would also decrease the potential involvement of these nations in fighting within the third world. In addition, it would free an enormous quantity of material and human resources for more constructive ends. And it would provide an opportunity for the growth of a spirit of community and the development of alternative institutions for resolving conflicts within and among nations. But this is getting far ahead of our story: We have yet to argue the case for the first move. Our case rests on two main points.

First, we argue that the alternative military policy which we propose is not only safe, but actually safer than the present policy, even if there is no corresponding, constructive change in other countries.

Second, we believe that the United States, the world's foremost military power and one of its largest and most prosperous nations, bears distinct responsibility to begin to wind down the excessive militarism of our times.

2
PRESENT U.S.
MILITARY POLICY

To describe U.S. military policy in clear, logical terms is to impute to it more coherence and control than it deserves. The policy is not constant but continually changing. It is formulated and implemented by a large and disparate group: civilians in the Department of Defense, military officers in planning positions, State Department and National Security Council officials, civil servants in the President's Office of Management and Budget, managers in the defense industries, and so on. Conflicting tendencies are produced by differing interests among the policymaking groups. Disparities arise between doctrine and action.

This chapter and later ones deal with these difficulties by looking at the *capabilities* of U.S. and other military forces. Potential uses of the forces are inferred from the situations in which they would be used cost-effectively. The study of capabilities is supplemented by official accounts of the purposes of the forces and by analysis of military exercises and of actual examples of use in war.

More detailed treatment of most of the following material is to be found in the later portions of the book.

DEFENSE, DETERRENCE, AND INTERVENTION

The main feature of the present U.S. military policy is the fact—too extraordinary to be believed by novices and too familiar to be discussed by experts—that there is no military threat to the United States against which the country can and does defend. No more than a few percent of the budget of the Department of Defense is spent on "national defense" in the narrow and literal sense of the phrase.

The explanation of this unknown fact (a gargantuan $100 billion case of "the Emperor's new clothes") is as follows. There is indeed a real military threat to the United States: This is the threat of an overwhelming, obliterative nuclear attack by the long-range bombers and missiles of the Soviet Union. It is the Soviet missile attack, in particular, which constitutes the danger. Against these missiles—which, in less than an hour's time, could destroy the entire U.S. urban population—*there is no means of defense.*

It must quickly be added that about 20 percent of the Defense Department budget is aimed at protecting the United States from a Soviet nuclear attack. However, it does so not by preparations to destroy the incoming missiles or defend the population, but by "deterrence." This means that the United States maintains its own intercontinental missiles and bombers which are intended to *deter* Soviet leaders from launching an attack *by threatening retaliation in kind.* These deterrent weapons, which absorb about one-fifth of the budget, are called "strategic nuclear forces."

The other 80 percent of the Defense budget is spent on traditional Army, Air Force, and Navy forces. These forces, with their nonnuclear weaponry—called "conventional" weaponry—are not aimed at defending the United States, for there is no conventional military threat to U.S. territory. Neither the Soviet Union nor any other country can mount a nonnuclear attack on this country. Potential enemy forces lack the transport aircraft and ships, the long-range tactical air cover, and the extended, protected supply lines which would be required for such an attack. The functions of the traditional forces which absorb most of U.S. military spending are not precisely spelled out in the defense budget, where they are called, vaguely, "general-purpose forces." The Annual Reports of the Secretary of Defense and other material presented to the Congress make clear, however, that the functions of the general-purpose forces include the defense of allies, intervention in foreign conflicts, and support of U.S. overseas interests—not direct defense of U.S. territory.

OFFENSIVE STRATEGIC FORCES: WEAPONS FOR DETERRENCE

In military terminology, the word "strategic" refers to the vital interests of countries and to forces which protect, threaten, or attack homelands. Formally, there are two categories of "strategic forces" in the U.S. defense budget: *Offensive*, deterrent weapons, designed to protect the U.S. homeland by threatening the homeland of potential attackers (the Soviet Union); and *defensive* weapons. Almost all of the military funds spent on strategic forces go to the offensive weapons. For this reason, the phrase "strategic nuclear forces" or "strategic weapons" is generally used to refer to the offensive, nuclear-attack systems.

The United States maintains three kinds of forces for strategic attack: (1) jet bombers (about 400 B-52s and FB-111s); (2) land-based intercontinental ballistic missiles (ICBMs) kept in concrete-reinforced underground silos, from which they can be launched on a few minutes' notice (1,000 Minuteman and 54 Titan missiles); and (3) somewhat smaller missiles, based, 16 apiece, on 41 nuclear submarines, from which they can be launched submerged through deck hatches (656 Polaris and Poseidon missiles).

In addition to their main function of threatening retaliation for an attack on the United States, the strategic nuclear forces have a largely unrelated secondary function. This is to deter nuclear or conventional attacks on U.S. allies and inhibit other foreign actions opposed by the U.S. government. Thus, U.S. nuclear forces present the Soviet Union and any other potential enemies with the risk that a serious conflict involving the United States might escalate up through a conventional war to the point of a U.S. nuclear attack. The policy of deterring foreign acts other than an actual nuclear strike on the United States by threatening a U.S. nuclear strike in response to such acts is referred to as "extended deterrence."

Since 1950, the United States has supported the concept of extended deterrence by refusing to agree to an international pledge of "no first use" of nuclear weapons. The U.S. government has made it clear that it might indeed be the first to use nuclear weapons, doing so in response to secondary, nonstrategic provocations. If the United States actually were first to launch a nuclear attack in a situation of armed conflict in some other part of the world, and if the conflict involved the Soviet Union, this would put the U.S. population at risk of an exchange of U.S. and Soviet nuclear weapons directed at each other's homelands. In this way the extended deterrence function of the strategic weapons increases the risk of a nuclear attack on the U.S. population. The secondary function of protecting others is at odds with the primary function of protecting the U.S. population.

The United States long ago achieved the capability of destroying the Soviet urban population in a retaliatory strike. Virtually all production of U.S. strategic weapons during the last decade, as well as planned developments during the next, improve a different aspect of the U.S. strategic force: The capability to attack Soviet military targets in a "limited" nuclear exchange. They have done this by increasing the number and accuracy of the nuclear warheads and by permitting rapid changes in their targets.

This countermilitary (also called counterforce) capability is not needed for the primary, deterrent function of the strategic forces— the function of deterring a Soviet attack on U.S. cities by threatening retaliation on Soviet cities. The countermilitary improvements are useful, rather, for the secondary, extended deterrent role of the strategic forces. They strengthen this role by making more possible and more plausible the hypothetical situation in which the United States would use nuclear weapons first in a limited and calculated response to some objectionable move by the U.S.S.R. To be a believable deterrent, this must imply that the United States would be prepared to escalate slowly toward a holocaust from any given level of nuclear exchange. The U.S. capability to attack nonurban targets and to escalate slowly toward urban attack is intended to provide a stronger deterrent of actions other than a nuclear attack on the U.S.A., again at the price of increased risk to our population.

DEFENSIVE STRATEGIC FORCES: WEAPONS FOR DEFENSE

Due to the great speed of incoming intercontinental nuclear missiles (15,000 mph to 20,000 mph) and their enormous destructive power, there is no effective means of defense against them. The false sense of security offered by partial and inadequate antimissile systems is well recognized, as is the futility of an arms race involving the development of such systems and the production of larger and larger numbers of offensive weapons to overwhelm them. On this basis, the United States and the Soviet Union signed a treaty in 1972 limiting the deployment of such systems to no more than two sites. This was later reduced to one site, and in 1976 the United States shut down its single antimissile site, located at a missile base in North Dakota.

Although the United States does not have any antimissile defenses, it does have an antibomber air defense, comprised of a large number of radars and 330 long-range interceptor aircraft. This continental air defense system, which costs a few percent of the Defense Department budget, is the only component of U.S. military forces oriented strictly to the defense of U.S. territory. As a result of the lack of a plausible bomber threat to the United States, this one defensive component of U.S. military forces has been substantially reduced over the past fifteen years. The number of interceptor aircraft has been cut by half, control of most of the remaining interceptors has been turned over to the Air National Guard, and the complement of over 1,000 surface-to-air Nike-Hercules launchers, placed around the United States in the late 1950s, is all but phased out.

In addition to its light air defense, the United States has a modest coastal defense, provided by the U.S. Coast Guard. The Coast Guard budget—originally covered by the Treasury in the interest of catching smugglers—is now covered by the Department of Transportation, not the Department of Defense. Around 40 Coast Guard cutters and other large armed patrol vessels are usually not counted as U.S. combat ships in naval balance calculations, since they are not ever expected to be confronted with an attack by foreign warships in U.S. coastal waters.

GENERAL-PURPOSE FORCES

The military goals of U.S. general-purpose forces are particularly hard to pin down. These forces are intended to be used flexibly in a variety of contingencies and regions. Their size and structure are determined in part by tradition rather than current or future military needs. Moreover, due to lack of Congressional concern with the rationales for these forces, little or no attempt is made in the annual budget presentations to explain why the United States needs general-purpose forces of the present size, rather than, say, forces 50 percent smaller or 50 percent larger. Much study is needed to clarify the threats these forces guard against and the foreign forces they are intended to oppose. The information relevant to developing a sense of their purpose includes: (1) an analysis of the foreign countries with which the United States is allied under mutual defense agreements, and of the military forces of those countries and their potential opponents; (2) illustrative descriptions of potential uses of the forces made by defense officials; (3) naval patrols, permanent foreign deployments, and exercises; and (4) analysis of the implications of military performance factors (range, speed, endurance, invulnerability, and so on) designed into the weaponry.

The general-purpose forces comprise the regular Army, Navy,

and tactical Air Forces. Unlike the U.S. nuclear arsenal, these forces resemble the military forces used in World War II; and like U.S. troops in World War II, they are intended to fight across the Atlantic and across the Pacific.

There are five main components of the general-purpose forces:

(1) *Ground troops for land combat*: about one million men, organized into 16 Army divisions, 3 Marine divisions, and various independent units.

(2) *Land-based fighter and attack aircraft*: (*a*) to provide air cover for ground troops by fighting off enemy bombers and fighters and laying down advance fire over enemy troops and (*b*) to undertake independent bombing missions: about 1,800 Air Force tactical combat aircraft, organized into about 80 squadrons, plus about 350 Marine combat aircraft, organized into 25 squadrons.

(3) *Aircraft-carrier-based fighter and attack aircraft*: (*a*) to provide air cover and bombing missions in areas where friendly ground bases are not available and (*b*) to permit air attack of naval vessels at sea: 13 aircraft carriers, carrying a total of about 1,100 Navy combat aircraft.

(4) *Transport aircraft and amphibious assault and other transport ships*: for conveying Army and Marine ground troops, together with their weaponry and support equipment, to overseas "theaters of combat": 68 mammoth tank-transporting jet aircraft plus 234 other large military transport jets, providing immediate overseas airlift of about one-fifth division per day; 63 amphibious warfare ships capable of transporting one to two ground troop divisions for forced landings on hostile beaches; and a small number of other transport ships, supplemented by civilian reserve cargo aircraft and merchant marine reserve fleets.

(5) *Surface ship escorts and submarine and aircraft patrol forces*: (*a*) to defend the aircraft carriers, amphibious warfare ships, and supply convoys against aircraft and submarine attack at sea and (*b*) to search out enemy submarines over wide ocean areas: more than 160 cruisers, destroyers, and frigates; about 70 "hunter-killer" (antisubmarine) submarines, and about 200 long-range, land-based antisubmarine patrol aircraft.

Of the approximately 20 Army and Marine ground combat divisions, eleven and one-third are located in the continental United

MAJOR US FORCES AT HOME & OVERSEAS

	Far East	in the US	Europe & Mid-East
Army & Marine Divisions: heavy	🛡	🛡 5	🛡 5
light	1	9	⅓
Navy: Aircraft carriers	🚢 2	🚢 9	🚢 2
Major gun & missile ships	20	125	15
All Services: Combat aircraft	✈ 600	✈ 2600	✈ 1100

FIG. 2

The major U.S. land, sea, and air forces deployed at home and overseas about the beginning of 1978. In this summary we regard the U.S. zone as including Hawaii, Alaska, and the U.S.-held Pacific islands. The one Army division long in the Far East—Korea—is listed in the U.S. because it is scheduled to come home slowly, beginning in 1978. Strategic forces are not included.

The main weapons of surface warships are both guns and missiles of a wide variety. Usually these types are divided, though some ships carry both. We give the total here for all major types of surface ships.

The combat aircraft listed are the major armed types which support ground combat in all services.

This and many subsequent figures require counts of forces. We emphasize that in fact such counts are not simple. The forces and deployments are in dynamic change; much equipment is under repair or even conversion at any time; within one class of type there are often variations of design, or even of age, which make counting ambiguous. There are extra ships, tanks, or aircraft, say, in various states of storage; sometimes these are to be counted, sometimes not. Throughout this book we have tried to make realistic estimates of the active full-time forces. Rounding of totals often provides a better understanding than efforts at counting nose by nose in a complicated situation.

MANPOWER FY 1978, IV-3 and V-10 to V-23.

States, five and one-third are stationed in Europe
third are in and around the Pacific (one in South
the other two in Japan, Panama, Hawaii, and ꜰ
Several of the U.S.-based divisions have, in addi
based equipment, a complete second set of equ.
tioned" in Europe, so that the troops can be flown
the European-based troops at very short notice. ꜰ ɪɪɪs deployment
pattern suggests an orientation of the ground forces and associated
tactical air forces of about 50 percent to aid in the defense of
Western Europe, 10 percent for Asia (mainly Japan and South
Korea), and the remaining 40 percent freely available for combat in
either of those theaters or elsewhere.

In the past, the aircraft carriers and amphibious landing ships—
the Navy's "power-projection" forces—were included among the
forces that might take part in a major war with the Soviet Union in
Europe or Japan. During the last 10 to 15 years, however, these
very large and expensive naval vessels have become increasingly
vulnerable to disabling attack by missiles, both nuclear and
conventional. Though they retain some chance of use on the
periphery of major combat, it is now generally conceded that their
main utility lies in interventionary operations against much weaker
opponents, such as those in Vietnam. When the escorts and supply
ships accompanying the power-projection ships are taken into
account, it appears that about half of the Navy's general-purpose
spending and forces is useful mainly for third world interventions.
The other half of the Navy—ships to defend convoys and antisubmarine patrol forces—is oriented primarily toward a repetition
of a World War II-type conflict: resupplying U.S. ground forces
over the North Atlantic sea-lanes in the event of a protracted
nonnuclear war between East and West in Europe.

In the aggregate, roughly half of the general-purpose component
can be argued to have as its primary rationale assistance to the
defense of Western Europe in the event of a major nonnuclear war
involving the Soviet Union and its East European allies. The other
half—much of which could also be used in Europe—remains
oriented primarily toward combat in the Far East and toward
intervention in third world areas—Latin America, Africa, the
Middle East, and South Asia.

Since World War II, the United States has signed formal mutual
defense treaties, ratified by Congress, with 42 nations (Figure 3).
These nations are: 12 West European countries and Canada in the
North Atlantic Treaty Organization (NATO); 20 Latin American

US WORLD MILITARY ALLIANCES

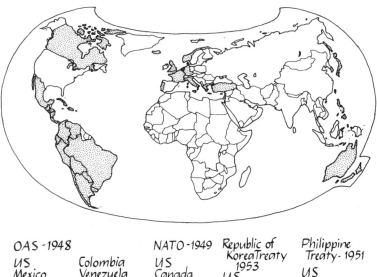

OAS -1948

US	Colombia
Mexico	Venezuela
Cuba	Ecuador
Haiti	Peru
Dominican	Brazil
Republic	Bolivia
Honduras	Paraguay
Guatemala	Chile
El Salvador	Argentina
Nicaragua	Uruguay
Costa Rica	Trinidad
Panama	&Tobago

NATO -1949

US
Canada
Iceland
Norway
UK
Netherlands
Denmark
Belgium
Luxembourg
Portugal
France
Italy
Greece
Turkey
W Germany

Republic of Korea Treaty 1953

US
Korea

Japanese Treaty-1960

US
Japan

Rep of China Treaty-1954

US
Taiwan

Philippine Treaty- 1951

US
Philippines

ANZUS Treaty - 1951

US
New Zealand
Australia

FIG. 3

Diplomacy can give legal expression to military goals abroad. Note the security treaties which link the United States formally worldwide to Europe, the Middle East, the Far East, and to the Americas. Such treaties are of subtle and shifting significance; for example the U.S.A. is hardly an ally now of the Republic of Cuba, which is, after all, still a full member of the Organization of American States. (In fact, we currently operate a naval base in Cuba.)

Department of State document, publication 8909, Sept. 1977.
The map is based on the armadillo projection by Erwin Raisz.

countries in the 1947 Inter-American Treaty (Rio Pact); Australia, New Zealand, Pakistan, the Philippines, and Thailand —all members, along with the United States, Britain, and France, of the 1954 South-East Asia Treaty Organization (SEATO), which was dissolved in 1977; and South Korea, Taiwan, and Japan, with whom the United States has bilateral mutual defense treaties. Following the dissolution of SEATO, the United States remains allied with Australia, New Zealand, and the Philippines through special treaties.

Of these formal U.S. military allies, the European nations, Japan, and South Korea are the main claimants to U.S. general-purpose force resources. These countries lie in close proximity to the Soviet Union and its massive ground forces and, in the case of the Far Eastern countries, in close proximity to the forces of China and North Korea. They also have extensive trade, cultural, and political ties with the United States. The requirement for direct U.S. military support in the Far East has been much diminished in recent years by the Sino-Soviet split and military confrontation, by the U.S.-Chinese rapprochement, and by the buildup of South Korean ground forces larger than those of North Korea. On the European front, the large size of Soviet ground forces remains an important potential military threat. Even here, however, the likelihood of a major World War II-type conventional war is extremely low, given the political and economic costs and benefits of war and peace to the two sides and the risk of escalation of any major conflict involving the U.S.A. and the U.S.S.R. to a nuclear holocaust.

Elsewhere in the world, the United States has a long-time defense commitment to Israel, with whom, however, it has no formal mutual defense treaty. The United States also has a growing mixture of military interests relating to the Arab states around the Persian Gulf. These countries supply most of the oil consumed in Western industrialized nations (almost all in the case of Japan). In addition, using their oil income, the Arab states and Iran have become major purchasers of U.S.-produced weaponry, creating a new region of financial importance and military concern.

It can be argued that the network of U.S. overseas interests and defense commitments is so large that the United States cannot afford to maintain separate general-purpose forces for the protection of each region of concern. Thus, instead of being bound strictly to certain geographic areas or to certain well-specified military threats and contingencies, the general-purpose forces can be described as flexible and mobile—available for easy movement from one area to another, to meet all the commitments listed above at a reasonable cost.

On the other hand, it can also be argued that for many of the countries the United States is committed to defend—in the extreme case, for all except Israel, to which the United States has no ground troop commitment—there is today and for the foreseeable future no likely threat to external attack or invasion. To the extent that this is the case, the function of U.S. general-purpose forces is not to meet defensive obligations to other nations: It is to permit the United States itself to play a policing or interventionary role in internal conflicts around the globe. This role has been illustrated in Vietnam as well as in the smaller-scale military operations in the Dominican Republic, Lebanon, and the Bay of Pigs.

Since the interventionary and defensive roles and rationales of the general-purpose forces vary in their importance to the individuals in the executive and legislative bodies responsible for forming and implementing U.S. policy, no exact assessment of the relative importance of each is possible. A clearer picture of the international position of U.S. general-purpose forces is, however, provided in the following chapters, in which U.S. forces are compared with those of other countries and set in the context of the military threats they might meet.

3

U.S. FORCES IN A
GLOBAL PERSPECTIVE

By most measures, the United States is the world's foremost military power. U.S. nuclear weapons are more numerous than those of the Soviet Union or any other country, and these weapons are generally more advanced technically. Indeed, the United States leads the world technically in nearly every significant aspect of military hardware, nonnuclear as well as nuclear. This technical proficiency contributes to the power of the traditional U.S. Army, Navy, and Air Force elements, but it is not the sole source of their strength. It is the combination of technical advantage with size and structure that makes American general-purpose forces uniquely powerful. These forces are capable of bringing nonnuclear power to bear in any part of the world. In this respect, the United States stands alone: It is the world's only truly global power.

In other respects, the United States shares military preeminence with the Soviet Union. Similar strengths in the military establishments of the two countries put them in a class apart from all other countries as the world's superpowers. Both the U.S.A. and the U.S.S.R. have thousands of nuclear warheads on intercontinental-range strategic missiles. They have, in addition, the world's most powerfully equipped land-combat forces, and each maintains an oceangoing navy several times larger than the navy of any other country. The traditional naval and land forces are equipped not only with conventional weaponry, but also with thousands of "tactical" nuclear weapons. (The tactical nuclear arms supplement the superpowers' strategic arsenals; they are intended mainly to destroy opposing conventional military forces at ranges of some tens or hundreds of miles. See Chapter 12.)

As demonstrated in Vietnam, America's extraordinary nuclear and conventional military potential does not insure automatic victory over second- or third-rank military powers. In fact, the history of British, French, and American involvement in colonial wars over the last two centuries shows that large, technically advanced military forces are often incapable of eradicating guerrilla movements. U.S. conventional military power does, nevertheless, permit the United States to intervene in disputes, crush small-scale hostilities, or engage in large-scale war in any part of the globe. The Soviet Union, in contrast, does not have the military allies or the logistical network needed to conduct major conventional wars in regions remote from its own territory (the Americas, Africa, and South Asia). With boundaries contiguous with 12 countries (Norway, Finland, Poland, Czechoslovakia, Hungary, Romania, Turkey, Iran, Afghanistan, China, Mongolia, and North Korea), the U.S.S.R. can, however, wage war in many parts of the world directly from home territory or via short overland or enclosed sea routes.

In reality, it is the Soviet Union's overland access to Europe, the Middle East, and the Far East, rather than any Soviet threat to the American continent, which challenges the United States militarily. Informed American analysts who are troubled by the Soviet military threat will readily grant that—except for the extremely unlikely event of an unforewarned U.S.–Soviet nuclear exchange—a war between the U.S.A. and U.S.S.R. alone, involving mainly or exclusively their own territories, is a scenario so implausible that it is irrelevant to military planning. Moreover, if the objective of U.S. military policy were confined to the single goal of deterring the unlikely sudden nuclear attack, the cost of the U.S. strategic nuclear weapons maintained solely for this purpose would probably come to no more than about 10 percent of the present U.S. military budget.

Today's military confrontation between the United States and the Soviet Union is, thus, very different from the territorially bound pre-World War I or pre-World War II arms race where the arms purchases of the foremost military powers were direct means of aggression and defense among them. The U.S.–Soviet military competition is, rather, a contest mediated by third countries. Each superpower threatens directly not so much the territorial integrity of the other as its global influence and the territory of its allies. This is particularly true from the American point of view. A Soviet defense planner might make a case that the United States, with its powerful Navy and long-range logistics, acting in concert with allies in Europe and the Far East, could undertake a conventional invasion of Soviet territory. No U.S. military analyst claims that the Soviet Union could do the same to this country.

Although a direct nuclear or nonnuclear war between the United

States and the Soviet Union is extremely unlikely, wars involving their respective military allies—wars which risk bringing in one or the other of the two superpowers—are much more likely and have in fact occurred during the nuclear era. Past examples include the Korean War, the Vietnam War, and, at a lower level of superpower involvement, the series of Arab-Israeli wars and the Soviet interventions in Hungary and Czechoslovakia. The Korean peninsula and the Middle East remain candidate areas for future conflicts involving one or the other of the superpowers—or, conceivably, both. In the European region, a variety of circumstances can be imagined which might lead more or less quickly to the involvement of one or both superpowers in a major conventional war: a change in government in Italy or Yugoslavia followed by a civil war; a substantial conventional arms increase or acquisition of nuclear weapons by West Germany, provoking some sort of preemptive Soviet military attack; or a major uprising in Eastern Europe aided by the West.

Hypothetical future conflicts involving third parties are the stuff that U.S. and Soviet military fears and suspicions are made of. Even in areas where the outcome is granted to be not vitally important to either superpower—and, ultimately, no foreign interest is worth the price of a nuclear holocaust—it can be argued that the mere existence of conflicting interests may eventually impinge on superpower security through some chain of escalation. This dire chain leads from armed hostilities involving neither superpower through major conventional war involving both, to a first use of short-range tactical nuclear weapons on the battlefield. This is then followed by limited use of long-range strategic weapons against military and industrial targets (with hundreds of thousands or millions of "collateral" civilian deaths) and, in the end, by mutually suicidal nuclear attacks on cities.

The links from U.S. and Soviet foreign interests to the conventional arms which protect those interests and to the tactical and strategic nuclear weapons which deter further challenges to them form the rationale for the comparison of U.S. and Soviet arms which follows. This survey provides a broad backdrop against which the needs for specific U.S. military force components are assessed in later parts of the book.

STRATEGIC NUCLEAR WEAPONS WORLDWIDE

The Soviet Union and the United States have by far the largest nuclear arsenals in the world, with 2,000 to 2,500 long-range missiles and bomber aircraft supported by each superpower (Figure 4). The Soviet Union has more missiles (about 2,250 to 1,700), but the U.S. missiles and bombers can deliver several times more independent nuclear bombs or "warheads" (about 9,000 to 3,000). Following the U.S. lead, the U.S.S.R. has recently begun to deploy new missiles with multiple independent warheads; and it is narrowing the payload disparity at a rate of about 1,000 warheads per year. The United States retains advantages, however, in the capabilities for command and control (particularly retargeting), accuracy, selection of weapon effects, and post-attack assessment. Moreover, due to the greater vulnerability of Soviet strategic bombers and submarines, a larger proportion of the U.S. arsenal than the Soviet would be likely to survive a first strike by the other side. Advantages of these kinds are believed by some defense planners to be useful in the event of limited nuclear exchanges. For this reason they are claimed to strengthen the extended deterrent value of the strategic forces. However, as suggested in Chapter 2, the close linking of a potential U.S.–Soviet urban nuclear exchange with deterrence of conventional wars overseas, by the intermediary of either tactical nuclear weapons or countermilitary strategic weapons, is a perspective which can increase rather than reduce the danger to the U.S. population. This point is discussed further in Chapter 8, which describes the dangers of the quest for nuclear advantage. Despite these dangers, both the United States and the Soviet Union are currently investing large sums of money to replace their current "generations" of strategic delivery vehicles with new versions which improve the countermilitary capabilities of both forces.

The only other countries with nuclear forces are China, France, and the United Kingdom. Each of these countries has about 100 strategic systems. Their aircraft and missile delivery systems are relatively vulnerable to preemptive attacks by U.S. or Soviet forces, are useful mainly against urban targets, and are generally confined to ranges of 1,500 to 2,000 miles. India, which has no announced active nuclear military force, recently test-exploded a nuclear device. It therefore has demonstrated its capability to construct nuclear weapons—weapons which could be delivered at some distance even by ordinary transport aircraft. Two

OVERVIEW OF WORLD STRATEGIC NUCLEAR FORCES

	Land Based Missiles	Undersea Missiles	Bombers
US	1054	656	380
USSR	1480	850	135
All Others	under 100	128	over 100

FIG. 4

This figure summarizes those nuclear forces which are intended for long-range use against enemy homelands: the strategic forces. Such weapons can be launched by land, sea, or air. (Chapters 5 through 7 take these matters up in detail.) Note the strength of the Big Two, all the more impressive because the ranges of the launch systems of Britain, China, and France are in fact generally well short of intercontinental range. The U.S.A. and the U.S.S.R. have many shorter-range and intermediate-range missiles in addition to their strategic systems. (See figure 18 for details.)

It is said that China is slowly developing submarine-launched missiles. India has tested one nuclear explosive device and has some fifty light bombers of intermediate range. Israel and the Republic of South Africa are rumored to possess nuclear weapons, so far untested. South Africa has a dozen light bombers, Israel none, but many effective fighter-bombers.

SIPRI YEARBOOK 1977, p. 4, and MILITARY BALANCE 1977–78, pp. 5, 8, 19, 53.

other countries which are rumored to have developed nuclear weapons, although they are not known to have tested any, are Israel and South Africa.

LAND-COMBAT FORCES WORLDWIDE

The land-combat forces of the United States and the Soviet Union—that is, ground troops and tactical air forces—are among

THE MILITARY SPECTRUM OF THE WORLD

Manpower in thousands of men

Spending in $ billions	less than 100	100-300	300-1000	1000-4000
over 100				USA USSR
20-100				
5-20	Saudi Arabia	Japan	WGermany France Britain Iran	China
2-5	Canada Australia Sweden Belgium	Israel Netherlands E Germany Nigeria Brazil	Egypt Italy Turkey Poland Spain	India
1-2	S Africa Switzerland Norway Denmark	Iraq Czechoslovakia Yugoslavia Argentina Indonesia Greece Syria	S Korea Taiwan	
less than 1	The rest of the world's nations, about 80 of them, fit here.	Thailand Romania Burma Ethiopia Cambodia Bulgaria Cuba Hungary Philippines	N Korea Vietnam Pakistan	

the largest in the world, numbering from one million to two million uniformed personnel. Only China and India match or surpass the superpowers in numbers of ground troops (around three million and one million, respectively) (Figure 5). The U.S. (Figure 6) and Soviet armies remain, however, far better and more heavily equipped than those of the two large but poor developing nations; and there is no comparison between the strong tactical air forces of the superpowers, on the one hand, and those of the two developing countries, on the other. The ground forces of the two most powerful U.S. allies in Europe, West Germany and France, numbering around half a million in each country, are also better equipped than those of China or India. Turkey, another U.S. NATO ally, supports a relatively large but poorly equipped army (350,000 men). The only other countries in the world with very large standing armies (300,000 to 500,000) are found among the large Asian countries long involved in local conflicts: Taiwan, Pakistan, North and South Korea, and Vietnam. In general, these and the other developing countries with large armies (100,000 to 300,000 men under arms) have proportionately quite small air forces, with no more than 50 or 100 or occasionally 150 combat aircraft. Among the richer industrialized countries which are medium-rank military powers (100,000 to 300,000 men under arms), larger air forces are common. The United Kingdom, France, West Germany, and Sweden each have around 500 modern combat aircraft, while Poland, Italy, and Australia have 200 to 300. Advanced aircraft and missiles, a larger ratio of tanks and other combat vehicles to soldiers and the support of related intelligence, communications and logistics services account for 10-fold and even 100-fold differences in the cost per soldier of land-combat forces of about the same size found in industrialized countries as against those of poorer developing nations. Presumably, the spending on modern equipment results in a significant increase—though perhaps not a porportionate increase—in the combat effectiveness of the forces.

FIG. 5

Here the armed forces of the world are arrayed according to two measures, money and men. Larger forces (ground troops plus air and naval manpower) appear on the right; higher military spending at the top. If a diagonal line is drawn from the lower left to the upper right, countries above that line spend more per man ($5,000 to $35,000 annually) and have better-trained forces, equipped with ships, aircraft, and missiles. Those below spend much less—in extreme cases $300 to $500 per man annually, providing subsistence rations and small arms.

MILITARY BALANCE 1977–78, throughout.

MANPOWER OF THE US ARMED SERVICES

	Active full-time	Reserves part-time
Army	790,000	590,000
Air Force	570,000	150,000
Navy	540,000	100,000
Marine Corps	190,000	30,000
TOTAL	2,100,000	900,000
Civilian Employees, all services	1,000,000	

FIG. 6

The military manpower of the wealthy United States is here listed by military service. The U.S. Air Force and Navy are relatively large compared to the U.S. Army ground forces. This indicates the fact that U.S. forces are highly technological and oriented toward the use of power far from our frontiers. (Compare Fig. 21.) U.S. active military "manpower" includes about six percent womanpower, in all services a growing proportion. The Reserves have about nine percent women.

MANPOWER FY 1978, II-5, XVI-1.

NAVAL FORCES WORLDWIDE

The United States has the world's largest and most powerful ocean-going naval forces, in both tonnage (range) and firepower (munitions delivery and reload). The 13 U.S. aircraft carriers, with their total complement of 1,100 combat aircraft, provide a unique capability for attack of overseas targets on land and at sea. Similarly, the U.S. amphibious assault fleet, capable of transporting 40,000 troops for forced landings on hostile coastal areas, is unparalleled both in the size of the amphibious attack that can be mounted and in its global range (provided by the large fuel capacity of the ships).

A large U.S. auxiliary fleet of oilers (for fuel transport) and supply and ammunition ships gives aircraft carriers, amphibious groups, and other naval combatants a unique capability to intervene rapidly in overseas conflicts. The auxiliary vessels are used to resupply warships at sea so that they can be kept on station for months at a time, close at hand in the event of conflicts. In the case of the carriers, the resupply capability also permits sustained offshore combat engagements, illustrated in the bombing of Vietnam. Other special features of the U.S. Navy, which contribute to its ability to engage in rapid intervention and support sustained combat abroad, are its extensive system of overseas bases; the nuclear power plants of a number of carriers and surface escorts, obviating the traditional naval constraint of limited fuel supplies; and the estimated effectiveness of U.S. antisubmarine warefare escorts and patrol forces, which are believed to be capable of riding out a Soviet campaign to block the sea lanes in a matter of weeks.

Like the United States, the Soviet Union has several hundred oceangoing warships: No other country has more than 75 (Figure 7). The Soviet Navy differs from the American in that it is designed not to project power but to obstruct it, with emphasis on coastal defense. The U.S.S.R. has over 600 coastal patrol craft and submarines which have no counterpart in the U.S. Navy. (The U.S. Coast Guard has some 100 patrol boats intended mainly for tariff control and rescue). At the other end of the naval spectrum, the U.S.S.R. has no medium-to-heavy attack-aircraft carriers. Its closest counterparts are the two Moskva-class large cruisers with antisubmarine helicopters and the three Kiev-class small carriers, which carry vulnerable, limited-range, vertical take-off light bombers. Unlike U.S. carriers, the Moskva- and Kiev-class ships are equipped only for antisubmarine warfare and light, close air cover, not for deep interdiction. Moreover, the Soviet Union has no long-range amphibious assault transport capability and only a modest oceangoing antisubmarine and antiaircraft escort and patrol capability. What the Soviet Navy does offer is an *antiship*-oriented submarine and surface fleet, including many smaller vessels (under 2,000 tons displacement) which are useful for operations in enclosed seas, such as the sea of Okhotsk and Japan in the East, and the Baltic, the Black Sea, and the Mediterranean in the West. Together with Soviet coastal patrol forces, these oceangoing and enclosed sea ships are intended mainly to block U.S. and other Western naval attacks—on the high seas if possible, closer in if necessary.

With some exceptions—the British and the French ranging in the

WORLD NAVAL FORCES

	Sea-going warships	Strategic & attack subs	Patrol subs	Coastal vessels
US	223	106	10	99
UK	71	13	18	20
France	54	5	22	40
Italy	22		10	10
Other NATO nations	140		60	175
USSR	258	192	230	450
Other Eastern Europe	6		15	270
China	21	2	65	860
India	28		8	25
Japan	46		15	15
Other Asia, Far East	106		27	334
The Americas	98		28	285
Africa, Middle East	20		5	235
TOTALS	1093	318	513	2818

FIG. 7

These columns report the numbers of the chief ships and craft of the navies of the world, sampled for a few important nations and regions. A few points: The powerful strategic and attack submarines, mostly nuclear, are found mainly in the fleets of the super powers, with a few held by the three next most powerful navies. The patrol subs, which are diesel-powered, torpedo-firing submarines, like those of World War II, are much more widely held. They run less than half the size of the attack subs, but may still be useful against shipping. Often they have a long range, though they require air for their engines, and cannot remain long in motion under water. The first column lists the familiar warships of various sizes, carrying guns, missiles (and rarely aircraft) in numbers. Gun- and missile-carrying ships are found in important numbers among the NATO allies

Atlantic and the Japanese in the Pacific—the oceangoing navies of other countries generally number no more than a dozen ships and are intended to supplement much larger coastal defense forces rather than to operate far from home territory.

TACTICAL NUCLEAR WEAPONS

Many of the short-range missiles (20 to 200 miles) and most of the tactical combat aircraft of the U.S. and Soviet conventional armies, navies, and air forces can be equipped with nuclear warheads as replacements for their ordinary high-explosive bombs. This provides the superpowers with several thousands of "tactical" nuclear delivery vehicles, which supplement their long-range strategic nuclear forces. These tactical nuclear weapons have no significant counterpart in other countries. (France has a few such missiles.)

SUPPORT STRUCTURE

For both the United States and the Soviet Union, the support structure—that is, the surveillance, reconnaissance, warning, command, control, communications, and supply forces—behind the combat forces is much larger than that of any other country. As one illustration, both the United States and the Soviet Union use a good number of very expensive specialized military reconnaissance and communications satellites to support their global operations. The United States has an advantage here, too, in the diversity, sophistication, and coverage of its satellite systems.

---

of the U.S. Different types of warships vary greatly in size, from 1,000 tons to 90,000 tons displacement. Many are old and poorly armed. The small coastal vessels are most widespread; the large numbers shown refer mainly to fast short-range craft with a crew of two dozen or so, a few light guns (or nowadays maybe a couple of missiles), useful in coastal defense, but around a tenth the weight of the smallest sea-going ships. U.S. craft of this type are in the Coast Guard; they are included in the last column.

JANE'S SHIPS 1977–78, p. 796 modified.

RESEARCH AND DEVELOPMENT

U.S. and Soviet military research and development (R & D) programs—employing several hundred thousand scientists and engineers in each country to advance scientific knowledge and develop new equipment for the purposes of war—are 10-fold to 100-fold larger than those in any other industrial country. The U.S. edge in innovation and technical advantage in military systems throughout the period since World War II is generally attributed not so much to the relative sizes of the U.S. and Soviet research programs, as to the environment for innovation in the two countries. Soviet civilian technology lags far behind that of the West, and Sbviet opportunities and incentive for publication, scientific travel, and exchange of information are much more limited. In addition, the Soviet economic structure, separating R & D institutions from production facilities and putting a premium on quantity production, discourages change in the production lines. What the Soviet Union has achieved with its massive research program is, in general, not the sort of breakthrough which often occurs in the United States, but rather a respectable response to the U.S. breakthroughs across the spectrum of small and large weaponry and supporting systems. Covering that entire spectrum is expensive and has remained beyond the reach of any third country.

DEFENSE SPENDING

Defense budgets are used as an indicator of overall military strength since they aggregate many disparate kinds of combat and support capability into a single rough measure. By this measure as well, the United States and the Soviet Union outdistance all other countries considerably. They spend 5 to 10 times more than the few second-rank military powers and even more—100-fold or greater—than the vast majority of the world's nations (Figure 5).

4

SECOND GREATEST
MILITARY POWER

The Union of Soviet Socialist Republics has the greatest land mass of any country in the world, stretching across half of Europe and all of Asia over an area two-and-one-half times that of the United States. The Soviet population, at about 255 million, is the third largest in the world, ahead of the United States (215 million), though behind China and India (about 950 million and 625 million, respectively). The Soviet economy, substantially converted from an agricultural to an industrial base since the Bolshevik revolution of 1917, remains significantly less productive per man than the more technologically advanced economies of the Western industrialized nations. As a result, a work force somewhat larger than that of the United States produces a gross national product (GNP) estimated to be only about half as great (around $850 billion for the U.S.S.R. in 1976, as against $1,700 billion for the U.S.A.). The Soviet GNP is still the second largest in the world, though Japan, now at $540 billion (the product of a population of about 110 million), and West Germany at $420 billion (population 62 million) are slowly closing the gap.

With its large population and economic and natural resource base, the Soviet Union has created a military establishment more powerful than that of any other country except the United States. As indicated in Chapter 3, the main features of Soviet military power are the large numbers of men under arms, the power of the tactical air and naval forces, and the destructive capacity of the intercontinental nuclear missile forces. These missile forces are 50 percent more numerous than those of the U.S.A. and 100-fold greater in explosive power than the forces of the second-rank nuclear powers (Britain, France, and China).

35

THE U.S.S.R. RANKS BEHIND THE U.S.A.

Despite the larger size of the Soviet Army and the larger numbers of Soviet strategic missiles, the Soviet military establishment must still be ranked as second to that of the United States, for several reasons touched on in the preceding chapter.

1.   Heavily equipped Soviet ground troops can be used only in areas contiguous to the U.S.S.R., since the Soviet Union has no outsized cargo aircraft to carry tanks and artillery, no long-range amphibious or other cargo ships designed to transport such equipment, and no overseas bases prestocked for land warfare. Only the U.S.A. has these means to conduct ground warfare on a global scale.

2.   With regard to the tactical air power which provides support and cover for ground troops, Soviet aircraft have more limited attack capability and are more defensively oriented than those of the U.S.A.

3.   For naval warfare, the United States has the powerful global offensive potential of its 13 large attack-aircraft carriers and its 63 transoceanic amphibious warfare ships, none of which has any real counterpart in the Soviet Union. The Soviet fleet, a defensively oriented navy, is composed almost entirely of surface ships and submarines intended to interdict (destroy or block) U.S. naval attacks and U.S. resupply of overseas land wars.

4.   Soviet intercontinental nuclear forces are limited in ways that may give them less certain and less massive retaliatory capability than that of the United States, following a hypothetical preemptive first strike by the opposing side. First, in comparison with the U.S. bombers, Soviet strategic bombers are only one-third as numerous. They have only half the range, and they have none of the modern aids for penetrating enemy air defenses as well as none of the extra resistance to the effects of nuclear explosions occurring at some distance. Soviet bombers also sit on unprotected airfields with virtually no chance of escaping a missile attack. On the U.S. side, 30 percent of the bombers are kept on 15-minute alert, probably long enough to assure escape. Second, in the case of the submarine-launched missiles, only about 10 percent of the Soviet strategic fleet is at sea at any given time, within range of targets and invulnerable to missile attack. The United States, in contrast, keeps over 50 percent of its strategic submarines at sea, where they are virtually undetectable. Moreover the superior antisubmarine warfare capabilities of

U.S. attack submarines, combined with the natural geographic advantage of the U.S.A. and disadvantage of the U.S.S.R. in access to wide-open versus narrow port exits to the oceans, make it likely that some or most of the few Soviet submarines at sea (but few if any of the many American submarines at sea) would be caught in an antisubmarine warfare net. All these factors tend to reduce the numbers of Soviet weapons likely to be available for retaliation to an extent that may offset or more than offset the larger numbers of Soviet missiles.

These isolated measures of military effectiveness do not demonstrate how prepared the United States is to deter and defend against Soviet aggression. Such proof must rest on more complex arguments, which take into account the allies that might be involved on either side of a potential conflict, the location and objectives of the war, incentives and disincentives to go to war, to reach a quick settlement, or to prolong the engagement, and so on. Such arguments are considered in detail later in the book.

In brief, our conclusions concerning the relative military strength of the United States and the Soviet Union when considered in this broader context are two-fold. First, with regard to strategic weapons—the only weapons that threaten the U.S. population directly—we argue in Part II of the book that disparities between Soviet forces and those of this country are in any case irrelevant to the single, acceptable function of U.S. strategic weapons. This function—deterring Soviet use or threats of use of nuclear weapons by holding out the possibility of retaliation in kind—is, in our view, adequately performed by the invulnerable U.S. fleet of strategic submarines no matter what the number, yield, accuracy, or other characteristics of Soviet strategic forces. Second, with regard to Soviet land warfare and tactical air and naval forces—which do not threaten the United States but which might be used in a war with China or with U.S. allies in Europe or the Far East—we conclude in Part III of the book that Soviet aggression against U.S. allies is adequately deterred and defended against by the current U.S. and allied forces.

ALARMIST CLAIMS ABOUT A SOVIET MILITARY BUILDUP

During the last several years, there have been a variety of claims about new trends in Soviet military forces—trends which, in contradistinction to the view expressed above, are argued to pose an unprecedented threat to this country. These claims have appeared in several different contexts: in statements by active and retired U.S. military officers, in Congressional budget testimony prepared by civilian officials of the U.S. Defense Department, and in analysis performed by the aides to some members of Congress, by certain consultants to the C.I.A., and by some private groups drawing on the expertise and reputation of prominent retired military and civilian officials. What these several sources share is a fear that present and future improvements in Soviet forces will result in a Soviet military stance in the early-to mid-1980s considerably more threatening to the United States than the present situation. A careful analysis of each of the trends under consideration has led us to the conclusion that the degree of change, or the implications of change, in specific instances is often exaggerated; and that the overall impact of current and likely future developments in Soviet forces does not warrant any new or special concern for the security of this country or its allies. The remainder of this chapter lays out the evidence for this conclusion.

Soviet developments which have aroused concern involve strategic forces, conventional forces, and the overall quantity of resources believed to be allocated to the military. In the strategic area, alarmists have argued that the Soviet Union is striving for, or has achieved, "military superiority" over the United States; that the U.S.S.R. is not only reaching for superiority but also preparing for nuclear war; and that the United States is now or may become incapable of carrying out an effective retaliatory nuclear strike in the event of a nuclear war.

In the conventional area, concern has been expressed about increases made over the last decade and allegedly still continuing in the following: the number of military personnel, the number of Army divisions, the numbers and quality of tanks and personnel carriers, the quality of the tactical air forces, and the numbers, armament, and "blue-water capability" (oceangoing range and deployments in distant seas) of warships. On occasion, concern about conventional improvements is expressed quite sharply. This applies, for example, to the patently exaggerated claim that "the U.S. Navy is no match for the Soviet Navy" and to the extreme view that West European defenses, bolstered by the U.S.A., have no chance of holding the line in a major land war with the Soviet Union in that area.

Alarm has also been expressed at the replacement, or possibly supplementing, of existing medium- and intermediate-range nuclear missiles aimed at Western Europe by new nuclear missiles. Attention has also been directed at improvements in the Soviet capability of defense against low-flying bombers—diminishing what has hitherto been an advantage for the West.

In the aggregate, the various Soviet improvement programs are estimated to have accounted for a steady growth (after allowing for inflation) of 3 percent per year in the resources absorbed by the military over the last decade—a period when, after the Vietnam peak, U.S. military spending has first fallen and subsequently slowly risen.

Dire warnings about Soviet military advances are not without precedent. In 1955 the American public was warned by defense analysts and government authorities of a "bomber gap," in which the United States was believed to be falling behind the Soviet Union. Five years later, in 1960, when the United States had produced four times as many long-range bombers and twice as many medium-range ones, the same individuals, now accompanied by the Presidential candidate John F. Kennedy, warned of a nonexistent "missile gap." In the late 1960s, there were comparable allegations of a Soviet thrust for a first-strike capability and for a formidable antimissile defense. All these past alarms, believed at the time by most of the public, were subsequently proved to be false. Moreover, throughout this period the United States maintained a preponderance of military power.

This history is repeating itself in most of the alarmist claims described above. Most of the claims are based on limited, simplistic, and partial numerical comparisons of which country is ahead or behind with respect to a particular facet of military capability (or bean-counting, in military jargon). Frequently, measures which are only slightly more sophisticated, but nevertheless quite revealing, are ignored—for example, rates of change, the timing of similar programs on the two sides, questions of "mirror-imaging," and which of the two countries is more provocative when viewed from the point of view of the other. Perhaps most important, new programs are identified as standard-practice "modernization" when undertaken in the West but described as "up-grading," "buildups," or "major advances" when undertaken in the Soviet Union. Soviet programs then appear provocative and novel, when actually they are exactly like comparable programs treated as normal and unprovocative in the West. Such changes are in fact not surprising—and should not be distressing to either side—since they can be and have been predicted well into the future. Plenty of time is given for response, if any is called for.

It is useful to consider a couple of examples of Soviet "modernization." The first generation of Soviet land-based strategic missiles, originally put in the field from 1965 to 1975—five years after their U.S. counterparts—are being replaced over the 1976 to 1981 period by a second generation. These new missiles have multiple, independently targetable warheads. The Soviet program in some respects parallels U.S. missile replacements of our first-generation (liquid-fueled) types made during the mid- to late-1960s. In other respects their improvements parallel the U.S. MIRVing program conducted from 1970 to 1977. The Soviet Union has still a third generation of missiles under development and not yet in production. When and if they are produced, they would bring Soviet land-based missiles up to a par with, or possibly slightly in advance of, the present U.S. land-based missiles. These Soviet advances have been predictable since they were first begun on the United States' side a decade ago. Similar parallels in the submarine-based missile programs on the two sides—with the Soviet Union here too lagging behind the United States by 5 to 10 years—are to be seen. In conventional weaponry, the Soviet Union, like Western countries, is replacing older tanks and personnel carriers and tactical aircraft with newer and more capable ones. These programs considered alone tell us little or nothing about the relative military capabilities on the two sides: Western countries, too, undertake modernization programs.

In the complex area of analysis and interpretation—as opposed to bean-counting—it is easy to underestimate or to exaggerate the import of possible future developments. Even the U.S. Joint Chiefs of Staff have recognized the difficulties in interpreting the so-called Soviet threat in their 1978 annual report to Congress: "A range of perceptions exists in the U.S. intelligence community on Soviet capabilities and intentions."

Superiority in strategic weapons and nuclear war-fighting.

Both the United States and the Soviet Union have the capability of inflicting unacceptable damage upon the other in a retaliatory nuclear strike. The U.S. Joint Chiefs of Staff recognized this in a 1977 report to Congress which stated that they do "not agree that the Soviet Union has achieved military superiority over the United

States"; and that, on the contrary, "the U.S. strategic forces can attain the damage criteria prescribed and are considered sufficient to achieve U.S. objectives today."

As noted earlier, the Soviet Union is in the process of modernizing its strategic nuclear force. The new Soviet missiles, with multiple warheads and greater accuracy, have a greater chance than the old Soviet ICBMs of destroying U.S. land-based missiles. Thus, they do place the U.S. ICBMs at greater risk. In this sense the Soviets are reducing the potential effectiveness of one arm of the U.S. strategic force. This has little or no military significance, however, since there is no present foreseeable threat to the 5,000 independently targetable nuclear warheads based on U.S. strategic submarines. The 160 warheads on just one of these 40 strategic submarines could destroy millions of people in all major Soviet cities—and none of the two dozen U.S. submarines constantly at sea is in serious danger! This fact will be addressed further in depth in Part II of the book.

Additional issues are raised by some Western analysts in discussing the nuclear balance. One of these is the magnitude of Soviet civil defense preparations. It is difficult to see much value in these discussions except to emphasize that both the Russians and Americans are quite dubious about the effectiveness of civil defense preparations. In the face of nuclear attack, it seems questionable, perhaps incredible, to imagine widespread and orderly evacuation of 100 million citizens from urban areas. And even if one assumes the dubious possibility of such rational action under threat of attack, what would the populace do to support itself in the wasteland of irradiated cinder and rubble?

In short, there is little foundation for allegations that the nuclear deterrent at the heart of U.S. military power is being eroded today or will be eroded over the next 10 to 20 years.

Troops.

In a strictly numerical comparison, Soviet military manpower is 75 percent again as large as the U.S. active military. Such a simplistic comparison is misleading for many reasons. First, U.S. forces are now volunteer rather than drafted while Soviet forces are still drafted. The quality of U.S. forces can be assumed to be higher. Second, the United States and the Soviet Union are in very different geographic situations. They perceive varying threats as a result.

The Soviet Union now utilizes a sizable portion of its troops to guard the Sino–Soviet border and to maintain the internal order in Eastern Europe. It has many units manning air defense. Military rather than civilians in the U.S.S.R. carry out much support and even military production. It therefore seems unreasonable of U.S. officials to assume that the Soviets should reduce their troop level to U.S. numbers when the functions of the deployments are so divergent. This is illustrated by Figure 8, comparing U.S. and Soviet forces.

Conventional equipment.

The easiest way to compare such conventional weaponry as aircraft, ships, tanks, helicopters, or artillery is to examine numbers in inventory. This practice gives misleading or inconclusive results when important factors such as quality, type, deployment, and tactics are ignored. Some examples will help to illustrate this point. The Soviet Navy outnumbers the U.S. Navy by about 10 percent in oceangoing surface combat ships. The U.S. Navy, however, is still considered second-to-none in the world. How can this be? It is due to the plain fact that the U.S. Navy is far ahead in ship tonnage and ship days at sea. In other words, U.S. ships include a greater number of larger, oceangoing vessels whereas many Soviet vessels are smaller, coastal ships limited in range. In addition, U.S. ships can spend more days at sea due to their larger size, greater reliability, and U.S. facilities for underway replenishment.

A second example of conventional weapon disparities—and the complexity of interpreting comparisons—is the tactical aircraft. The total Soviet inventory is claimed to be one-third larger than the U.S. inventory. Yet most Soviet combat aircraft are considerably more limited in range, speed, weapons load, and other important characteristics. Similarly, the Soviet tank inventory is alleged to outnumber the United States over three to one, yet one by one Soviet tanks are judged considerably inferior to their U.S. and NATO counterparts. These examples will be discussed further in the chapter on the NATO–Warsaw Pact balance. Recent Congressional testimony has pointed out that even the current U.S. and NATO aircraft and tanks are equal—and often superior—to even the newest Soviet models just beginning deployment.

These few examples from nonnuclear weaponry are cited to show the difficulty of making clear-cut meaningful comparisons between very different weapons systems. The U.S. Defense Department has plenty of difficulty in deciding among a few competing prototypes of its own with their widely varying tradeoffs of costs and capabilities; one can imagine the complexity of comparing weapons systems which have seldom if ever been tested against each other.

Military spending.

The official Soviet defense budget was somewhat over 17 billion rubles for 1976; converted at the official ruble-dollar exchange rate, this comes to $25 billion, about one-fourth American defense spending. This amount appears to understate such spending and has therefore been refigured according to a variety of methods—anywhere from about $75 billion to $140 billion. This wide range of official Western estimates, calculated by private economists and by government officials both in the C.I.A. and elsewhere, is indicative of the difficulty in producing an accurate estimate of Soviet military spending. The quandary is the result of several problems: the great differences in cost and price structures between East and West; varying qualities of the technologies, and the efficiencies of market sectors; hidden costs and subsidies in the military sector, particularly in the Soviet Union; and the invariably unreliable ruble-dollar exchange rate. For these reasons, exact military spending comparisons between the United States and the Soviet Union cannot be definitively made.

Soviet intentions.

Perceived military threats are a function not only of identifiable military capabilities, based on estimated troop and weapon deployments, but also of intentions imputed to an opponent. Soviet military intentions are inferred by Western analysts in part from Soviet writings on military doctrine. Unfortunately, this source is limited in many ways—by the dearth of translated documents, the propagandistic quality of much Soviet military writing, and the biases of Western Sovietologists. Nevertheless, some American defense analysts feel sufficiently confident to impute warlike intentions—often uniquely warlike to the Soviet Union—just as many Soviet writers impute mirrorlike intentions to the United States.

FUNCTIONAL COMPARISON OF US & USSR MILITARY MANPOWER

	US	USSR
Strategic forces offensive defensive	85,000 15,000	400,000
		550,000
Coast Guard, border guard, internal security	100,000	500,000
Military in the Far East & Pacific	100,000	600,000
Military with civilian functions: railroads, farms, construction		700,000
Forces available for European confrontation	1,900,000	2,100,000
TOTAL	2,200,000	4,800,000

The most firm conclusion that can be drawn from a close reading of Soviet military doctrine is that it is ambiguous. On the one hand, Soviet military doctrine is based on Marxist–Leninist ideology, positing continual proletarian-bourgeois conflict. Until the mid-1950s this was interpreted by the Soviets to mean that war with the capitalist world was inevitable and would in fact be started by the capitalist countries' need for control of markets—imperialist expansion. Since the Khrushchev speech at the 1956 Twentieth Party Congress, however, this stance has been modified. No longer is war seen as inevitable since the advent of large quantities of nuclear weaponry. Like most Americans, the Soviets view nuclear war as mutually destructive and define it as the absolute last resort in international relations. It is to be avoided at all costs.

On the other hand, it is evident in contemporary Soviet military writings that with regard to *nonnuclear* war, the Marxist–Leninist theory of just and unjust war (a view found in St. Thomas Aquinas, too) is still followed. The Soviet Union will support what are perceived as wars of national liberation, anticapitalist, or anti-colonial wars—"just" wars; but antisocialist wars of so-called reactionary forces—"unjust" wars—will not be supported. It is apparent from this that the Soviet Union might continue to intervene in the third world wherever it deems such moves helpful. One must realize that with any reduction in U.S. interventionary capabilities, Soviet interventions may increase. The United States learned a difficult lesson in Vietnam; the Soviet Union has also encountered difficulties in dealing with the third world, for example, in Egypt and the Horn of Africa in the 1970s. It is therefore certainly possible that the noninterventionary military policy

---

FIG. 8

The military manpower of two countries so different in geography and history as are the U.S.A. and the U.S.S.R. cannot be fully compared by a simple head count. Here is given one functional comparison based on studies by Congressional and Pentagon agencies. The U.S. has no border with an estranged great power, a much smaller concern with air defense, a tendency to employ civilians, rather than military personnel as "civilians in uniform," and a powerful offensive strategic force manned by far fewer people.

Representative Les Aspin, *Are the Russians Really Coming?*, Council on National Priorities and Resources, Washington: 1976. A brochure, with commentary from the DIA reported in *The New York Times*, July 20, 1976.

recommended for the United States in this book will eventually be adopted by other superpowers.

Soviet intentions are, like all national purposes, complex; they must therefore be analyzed with caution from the expressions of military doctrine and Marxist–Leninist ideology. Soviet military writings emphasize the necessity to be constantly vigilant and fully prepared for war; but they also emphasize the absolute necessity for arms control and disarmament. Small, conventional wars of national liberation will continue to find Soviet support at some scale, but any major war with the potential for nuclear confrontation will be avoided.

SOME TENTATIVE CONCLUSIONS

The Soviet Union is a superpower—militarily, politically, and economically second only to the United States. It is as such the major rival to the United States for world influence, and it represents the primary driving force rationalizing American defense spending. It is therefore of primary importance in any analysis of American national security and foreign policy.

Recent allegations of a drastic Soviet military buildup and a "Soviet threat," are, we hold, grossly exaggerated and often based on limited analyses and data. Simple numerical comparisons of troops, expenditures, and weapons inventories present misleading and oftentimes false conclusions. Proper analysis of the Soviet–American military balance is complex and permits much less than firm, exact conclusions.

Any examination of the Soviet–American balance shows up a mirrorlike sameness of perceptions. The U.S. pictures the Soviet people as aggressive, while it assumes peaceful intentions for itself; similarly, the Soviet Union describes the United States as warmongering, but sees itself as peaceloving. The truth, in all probability, lies somewhere between the two extremes. U.S. military forces have engaged in wars far from our own country; communist revolutions and wars of national liberation have been aided by the Soviet Union far from its borders. Serious consideration of the continuing arms development by both superpowers might lead anyone to question the peaceful intentions of either side.

The Soviet military threat is real to the United States, just as the American threat is real to the Soviet Union. Both sides should be careful, however, not to dramatize those structural threats, nor to evoke an undue overreaction; propagandistic statements about warlike intentions and false allegations of clear "superiority" cannot serve either country.

The Soviet–American relationship is both a cooperative and competitive one. It is competitive in the sense of continued socialist–capitalist conflict. Both countries have experienced years of cold war confrontation; this has sometimes resulted in coming to the brink of war—in Berlin in 1961 and Cuba in 1962. It has, however, more often than not involved less direct ideological competition through third-party intermediaries.

The relationship is cooperative in the sense of a mutual recognition of interdependence in the military and in other spheres. For example, between 1972 and 1974 the two countries signed 11 bilateral agreements for international exchange in such fields as health, energy, environmental protection, housing, and transportation. Since 1959 the United States and the Soviet Union have also signed a number of bilateral and multilateral arms control agreements. This leads one to believe that the Russians and Americans are coming slowly to recognize their mutual and overlapping interests, while continuing their ideological competition.

The Soviet Union is a good example of a country recently emerged as a world power; it should therefore come as no surprise that it desires to maintain—or even to expand—its world influence. This is not at all unlike the past 75 years of U.S. growth to power.

It is not our purpose here to analyze how the Soviet Union might better organize its forces—that would be the subject of quite another book. Yet we remark that the United States has the opportunity to encourage the Soviet Union to follow a reasonable example. U.S. and Soviet representatives are talking already in 11 or more arms control forums on such subjects as strategic weaponry, nuclear test bans, demilitarization of the Indian Ocean, European force reductions, chemical weapons, nonproliferation of nuclear weapons, and other important issues. The U.S. should continue to support these forums. Neither the United States nor the Soviet Union can want a superpower confrontation; yet both countries maintain enormous military arsenals, now more than adequate for national defense. The more both sides can be convinced of the necessity for unilateral and negotiated arms reductions, the safer their relationship will become, and the safer the whole world.

At the beginning of the twentieth century the United States did not rank militarily even among the first half-dozen major powers. We are now concerned about the second greatest power. As long as nations strive to be first, there will always be some power that is second; history tells us that the U.S. and the U.S.S.R. will not occupy these positions forever. It is our task to conduct our national affairs so that they do not jeopardize the peoples of both our nations, or of the world.

5

RELATING FORCES, BUDGETS, AND POLICY GOALS

This chapter sets forth the general objectives which, in our view, are acceptable goals of U.S. military policy. It also outlines the force-planning criteria which we attempted to follow in selecting military forces to meet these goals. It then shows how force levels are related to military spending.

POLICY GOALS AND FORCE CRITERIA

Initially, we assumed simply that the aim of U.S. military policy should be defense. In designing military forces to meet this goal, we followed three rules:

 (1) Assess the war-making potential and incentives of the United States and other countries—both allies and opponents—in a comprehensive and realistic way, rather than in partial studies of selected components. Partial studies tend to ignore overall intentions and capabilities and make "worst-case" assumptions about future military and political developments.

 (2) Maintain military forces which are as economical and spare as possible without jeopardizing their adequacy for their assigned functions.

 (3) Introduce any major changes in the present U.S. military posture in a manner that is sharpest where the need for change is most pressing and more gradual where the need is less urgent, so that the overall effect is a clear yet measured shift in policy.

Although we continued to use the "defense" requirement as our guiding objective throughout the study, we gradually came to the conclusion that it would be both unrealistic and unwise to recommend that in one stroke U.S. military policy should be reoriented so as to be confined exclusively to the territorial defense of the United States. Such a purely defensive policy would probably cost no more than a few percent of our present military spending. Both domestically and internationally, the change would be revolutionary. We believe that it is important not to propose an immediate change in the size of the U.S. military establishment and budget on a scale which could reasonably be expected to have disruptive impact on international relations. This might cause upheavals or increased tension in other countries and result in increased militarization and greater likelihood of war than exists today.

For this reason, we have not proposed that the United States move in one step to some sort of ultimate objective; nor have we thought it useful or necessary to debate—at this stage—whether the ultimate objective should be a small national defense force, a U.N. peacekeeping force, or some other military or nonmilitary structure.

Instead, we propose to maintain the most important, traditional U.S. defense commitments to other countries and to retain the portion of the general-purpose forces needed to meet these commitments. Similarly, we have not recommended the abolition of all U.S. nuclear weapons, even though these weapons are not defensive but deterrent, offensive armaments. A strong case can be made that keeping nuclear weapons invites attack more than it inhibits one. In fact, nonnuclear countries may be more secure from nuclear attack or blackmail than the nuclear powers. Still, we propose to retain a U.S. offensive nuclear force whose destructive potential, though small by present standards, remains cataclysmic.

The policy and military force structure we propose is, nevertheless, one of substantial retrenchment. It represents a major cutback from the fearfully large nuclear arsenal and a withdrawal from the dangerously trigger-ready global patrol currently maintained by the United States to a military posture that is clearly and unequivocally defensive. In our view, this is the largest and most dramatic change that the U.S. government can reasonably be expected to undertake in the space of a decade. It is a change with radical implications for the demilitarization of U.S. foreign policy.

The policy we propose is more defensively oriented than the present policy in the following respects. First, we eliminate the nonnuclear general-purpose forces which are suited more to intervening in conflicts against weaker countries than to countering aggression by the Soviet Union. This applies particularly to the lightly equipped land-combat divisions and to the naval aircraft carriers. Second, we draw down forces which exceed defense needs in regions where there is a plausible threat of Soviet aggression. Such surplus forces contribute to the likelihood or at least the fear that U.S. military power might be used aggressively. The U.S. contribution to the military balance between NATO and the Warsaw Pact in Europe is a case in point. We recommend a posture that is fully adequate for defense but less suitable for offensive operations than the present force structure.

Third, with regard to strategic nuclear forces, our proposals reduce the vast abundance of U.S. offensive weaponry. The present U.S. nuclear arsenal threatens to damage the Soviet Union and surrounding lands beyond any meaningful or defensible requirement of warfare—and far beyond the destruction we judge adequate for deterring a deliberate Soviet nuclear attack on the United States. Moreover, the extraordinarily large size, diversity, complexity, and sophistication of present and planned U.S. nuclear forces may create an incentive for U.S. leaders to blackmail the U.S.S.R.—to gain advantage in a crisis—with the threat of conducting a limited nuclear attack so precise and so finely controlled that the U.S.S.R. could not respond in kind. To inhibit adventurous U.S. behavior, it is important to eliminate the sort of surplus that permits serious thought about deliberate first use of nuclear weapons.

A smaller and more defensive U.S. military establishment would make it necessary for the government to give the American people a clearer signal when the United States became involved in an overseas war. This would make it more likely that involvement in such wars, if it occurred, would actually be supported by a broad popular consensus and would be, thus, in the *national* interest. It would also reduce the financial burden and the corrosive political influence of an economy and a job market which is nearly 10 percent military. Equally important, cutting out much of the excess in U.S. military capability would make the defensive orientation clear and unmistakable to foreign audiences, removing excuses for excess militarization elsewhere—particularly in the Soviet Union

The old saying that "the best defense is a good offense" may be a useful rule of thumb for tactical maneuvers in the midst of a war, but it is clearly not the way to a peaceful world and a demilitarized society.

DEFENSE SPENDING

By program
FY 78 $billions

1. Strategic Forces	10.8
2. General Purpose Forces	42.2
3. Intelligence & Communications	8.1
4. Airlift & Sealift	1.6
5. Guard & Reserve	6.9
6. Research & Development	10.8
7. Maintenance	12.0
8a. Training, Medical & other Personnel	15.3
8b. Retired Pay	9.1
9. Administration	2.3
10. Support of other Nations	1.3
TOTAL	$120.4

Programs grouped by activity

Combat Forces 1. Strategic 2. General Purpose 4. Airlift & Sealift 5. Guard & Reserve: 60%	58.7
Specialized Support Activities 3. Intelligence & Comm. 5. Guard & Reserve: 40% 6. Research & Development	21.7
Basic Support Activities 7. Maintenance 8a. Training etc. 8b. Retired Pay 9. Administration	38.7
10. Other Nations	1.3
TOTAL	$120.4

RELATING MILITARY FORCES TO THE MILITARY BUDGET

At present, the appropriations made by Congress for the Defense Department shed virtually no light on U.S. policy goals. Funds are appropriated not in the name of such military functions as "offensive, retaliatory nuclear forces to deter nuclear attack on the U.S.A.," "offensive nuclear forces to deter conventional attack on allies," "nonnuclear aid to defend Western Europe against aggression by the U.S.S.R.," and the like. Instead, funds are voted for accounting-type categories: "pay of military manpower," "procurement of weapons systems," "military construction," and so on.

In an effort to make the annual Congressional review process more sensitive to the political commitments implicit in each year's Defense budget, former Secretary of Defense Robert McNamara introduced a supplementary accounting system, which regroups Defense appropriations according to the 10 programs listed in Figure 9. The program accounting system aggregates U.S. *combat forces* into the two main kinds: the strategic forces (Program 1) and the general-purpose forces (Program 2). The transport aircraft and ships which make it possible for the general-purpose forces to fight in other parts of the world are separately listed under "Airlift and sealift" (Program 4), but these vehicles actually form part of the general-purpose component. Except for military aid to foreign

---

FIG. 9

The two bars show the Department of Defense budget for FY 1978, divided into military programs. There are ten programs, listed in the official order in the left bar. (These program numbers are central to the DOD accounting system.) In the right bar, we simply regroup the same programs by their general nature: the combat forces themselves, and special and basic support. The direct costs of strategic nuclear forces comprise less than 10 percent of the annual budget, according to this breakdown; the conventional forces comprise over one-third.

It should be noted that the DOD budget does not include the yearly expenditures (made now by the Department of Energy) for nuclear weapons and naval nuclear fuel: a sum of $2.2 billion in this year. It also does not include $2.6 billion expended on civil functions (like river and harbor works) by the Corps of Engineers, and $18 billion spent by the Veteran's Administration. (The fiscal year FY 1978 opened October 1, 1977, and closed at the end of September 1978. The present definition embodies a recent change; up to July 1976, the government's fiscal year ran from July through June.)

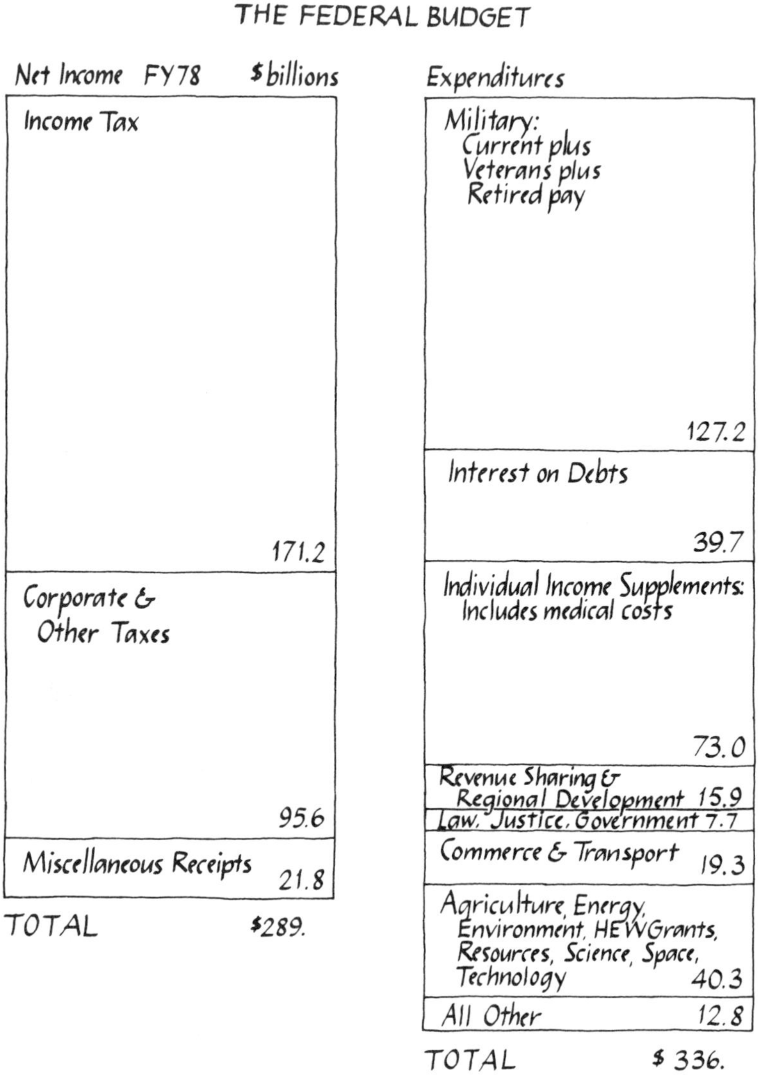

THE FEDERAL BUDGET

Net Income FY78 $billions   Expenditures

Income Tax
171.2
Corporate & Other Taxes
95.6
Miscellaneous Receipts   21.8

TOTAL $289.

Military: Current plus Veterans plus Retired pay
127.2
Interest on Debts
39.7
Individual Income Supplements: Includes medical costs
73.0
Revenue Sharing & Regional Development 15.9
Law, Justice, Government 7.7
Commerce & Transport   19.3
Agriculture, Energy, Environment, HEW Grants, Resources, Science, Space, Technology   40.3
All Other   12.8

TOTAL $ 336.

FIG. 10

These two bars graph the income and the expenditures of the Federal Government in the FY 1978 budget. The income bar is shorter than the expenditure bar, showing a deficit of $47 billion. In this accounting the direct transfer from the nations, the remaining items are all *supporting activities*, which back up the combat "teeth" of the strategic and general-purpose forces. For this reason, the defense and foreign policy objectives of U.S. military forces are implicit mainly in an analysis of the first two items alone—the strategic and general-purpose forces. Parts II and III of this book analyze these forces and their rationales in detail and recommend a number of changes in them.

Part IV of the book looks closely at the specialized support activities and the matter of military aid. Policy decisions in several special areas—the support of intelligence (Program 3), the mainte-nance of reserve forces (Program 5), and the volume of research and experimental work to develop new weapons (Program 6)—do not follow automatically from choices about the combat components of the forces, but must be separately considered. Each is given a separate chapter in Part IV. In contrast, the basic support activity—that is, the supply and maintenance of equipment (Program 7), the training and medical care of manpower (Program 8), and the administration of the forces (Program 9)—naturally does vary with the size of the combat forces. Apart from the irreducible minimum of such support needed to train and equip combat forces of a given size, decisions made in programs 7 to 9 which might have a noticeable impact on the overall military budget concern internal management and efficiency: There are few political issues with foreign policy implications. In Chapter 16 of Part IV these three programs are discussed only briefly.

Somewhat surprisingly, the pay of retired military personnel is included in the budget of the Defense Department, along with the expenses of the active military forces. In the program accounting system, it is included under Program 8, as a type of personnel expense. In our tables and discussion we have listed the item separately, however. The reductions in active military forces which

---

social security paid to social security recipients is excluded; the figures include only the additional net contribution made by the taxpayers, not that stored in trust from previous social security deposits. A rough summary of expenditures can be made simply: military, a little under 40 percent; income supplements to individuals and other government bodies, 25 percent; all other operations of government, 25 percent; and interest payments 10 percent.

FIGS. 9 and 10

US GOVT BUDGET FY 1978.

we propose will result in an increase in retired military personnel
and an increase in this budget item. This item has in any case been
rising rapidly in recent years as the 20-year turnover cycle
associated with the large standing army maintained since 1950 has
begun to show its full effect, and as retirement compensation rates
have been increased by the Defense Department and the Congress.
Retirement costs are covered in Chapter 18 of Part IV of the book.

Treated in Part V is the impact of our recommendations on the
budget on civilian jobs in the defense industries and, more broadly,
on the national economy as a whole. In Chapter 19 we calculate the
total budget cost of the combat forces and support activities which
we recommend, providing an overall alternative budget for the
Defense Department. In Chapter 20 we show how, with proper
planning, the economy can be strengthened by a cut in defense
spending, as the financial and human resources currently devoted to
the nonproductive "insurance" goods and services of the military are
turned to health, education, and other productive sectors. Part V
concludes in Chapter 21, a summation, with a short statement on the
expected impact of our proposals on this country and on the
international environment.

THE YEARLY OUTLAY ON THE MILITARY

We finish this introductory section of the book by totaling up how
much the United States spends each year on military purposes. The
Department of Defense is only one, albeit the largest, recipient of
military dollars. Several other agencies which we do not consider in
detail in this book also play some role. Most important, the nuclear
materials and actual nuclear warheads deployed by the Defense
Department—though not the aircraft and missiles on which they are
deployed—are financed and manufactured by the Department of
Energy (DOE) (formerly the Energy Research and Development
Administration, ERDA, and before that the Atomic Energy Commission, AEC). The purpose of this separation of responsibilities
was originally to insure civilian control over nuclear weapon policy:
In fact, DOE does little more than develop and produce warheads to
Defense Department order. The DOE budget for 1978 includes
about $2 billion for military programs—a small amount set against
the $120 billion budget of the Department of Defense. The DOE
military expenses are included under outlays for "National Security"

in the functional breakdown of the overall Federal budget issued by the President each year.

Two agencies with some military-related activities which are not included under the National security function are the Veterans Administration (VA) and the National Aeronautics and Space Administration (NASA). For the Veterans Administration, about $18 billion is budgeted for education, health care, other special benefits, and direct payments to veterans and survivors of participants in the Spanish–American War, World War I, World War II, the Korean War, and the Vietnam War. Unlike the retirement item in the Defense Budget, which compensates men with 20 years or more service in the Defense Department, the VA serves actual war veterans and other military personnel most of whom have been in service only a few years. In 1978 NASA is scheduled to spend $4 billion, of which $0.4 billion is for advanced aeronautical research and $2 billion for the space shuttle. Both research areas have a good chance of eventually yielding military applications, and indeed the shuttle was planned to allow certain military launches, in the expectation that the Air Force would defray some costs. Military technology today is so wide-ranging that much research and development, whatever its original aim and whoever its original sponsor, becomes useful militarily. To be sure, that flow of novelty is two-way: The jet airplane, the computer, and fusion laser development, to name only a few examples, embody military R & D in important civil use. We shall not try to disentangle this billion-dollar web.

Other military-related items in the national budget are the "civil functions" of the Department of Defense and a portion of the interest on the national debt. The Defense Department's civil functions, listed and financed in the federal budget separately from its "military functions," are almost exclusively dam, harbor, and other earth-moving activities carried out in the continental United States by the Army Corps of Engineers. The cost of these activities comes to $2.6 billion in 1978. Some of the interest on the national debt is sometimes referred to as an indirect military expense because part of the $700 billion debt—on which interest of about $40 billion will be paid in 1978—has been incurred through the sale of bonds to support wartime spending.

A final note is in order about budget support of the Central Intelligence Agency, C.I.A., which is nowhere shown in the published budget documents, and about other concealed military-related expenditures (such as for the National Security Agency, the eavesdroppers). The bulk of C.I.A. expenditures are believed to be included in the Defense Department budget, under the headings for intelligence and communications (Program 3) and research and development (Program 6), and counted within those programs with Air Force and Defense Agency funds. This would be the appropriate place for C.I.A. funds and there are, in fact, several large unexplained items in those accounts. It is possible, however, that some additional parts of the C.I.A. budget are financed through the accounts of other government agencies not mentioned hitherto—for example, in funds for science and technology, international affairs, commerce and transport, or funds appropriated to the President. Allowing for both possibilities, we estimate that the funds that could be conveniently concealed come, in all, to not more than $1 billion outside Defense Department accounts. plus several billion dollars within the Defense Department accounts—in the latter case, mainly in accounts labeled and analyzed by us, appropriately, as intelligence and communications.

If *all* the known activities with some military connection past and present were added together—that is, the Defense Department's military and civil functions, ERDA military programs, NASA military-related programs, the Veterans Administration, and interest on the national debt (not all of which is wartime debt)—the total comes to $175 billion in actual outlays budgeted for 1978. That is about half of budgeted federal outlays *excluding* social security and unemployment insurance payments, or about 40 percent of the total if social insurance transactions are *included* with other federal programs. Outlays for current military forces, including the Department of Energy nuclear weapons work but excluding retirement, veterans' benefits, and interest on war debt, come to $103 billion, just one-third of the federal budget, excluding social security and unemployment insurance, or about one-quarter of the budget including these transactions (see Figure 10), and larger than any other category there listed.

PART II
STRATEGIC
NUCLEAR FORCES

6

FOUNDATIONS OF
NUCLEAR DETERRENCE

Strategic nuclear weaponry is a sword of Damocles—to paraphrase the late President John F. Kennedy—which has been swinging over the world's populations for more than three decades. Even though the U.S. strategic nuclear force accounts for a relatively small part of the defense budget in dollars—about 20 percent—it carries a destructive potential so awesome that it is difficult to assess. Yet some understanding of the physical effects of nuclear weapons, of the means of placing them on intended targets, and of the logic of this unprecedented kind of warfare is necessary for discussion of policy.

In this chapter and the next two, we will confine our attention to the overarching problem of strategic nuclear weapons. A weapon is called strategic when it is intended for and likely to be able to reach targets far from any battlefield, deep inside the territory of a potential opponent. The actual physical effects, however, are virtually the same as those of "tactical" nuclear weapons which are oriented toward use against nearer battlefield targets up to some hundreds of miles away. The difference is more in the range of the vehicles which bring the weapons to the target and less in the nature of the weapons themselves. Strategic targets include some targets, such as great cities, which are of very large area; thus, strategic nuclear weapons sometimes possess greater explosive strength.

At present the strategic nuclear arsenal of the United States consists of about 9,000 nuclear warheads mounted on a so-called triad of long-range delivery systems: (1) the Air Force's intercontinental ballistic missiles—called ICBMs—based on land in buried silos; (2) ballistic missiles based in and launched from the Navy's submerged oceangoing submarines at sea—SLBMs (for submarine-launched ballistic missile); and (3) gravity bombs and short-range missiles (now a few hundred miles in range) carried on long-range

Air Force jet bomber aircraft. In addition, the United States has at home, at sea, and in Korea, Britain, and West Germany some 22,000 nuclear warheads (Army, Navy, and Air Force) which are called "tactical" because they are not intended to be carried more than a few hundred miles to a target. (The distinct but related question of tactical nuclear weapons is deferred to Chapter 12.) This makes a total U.S. strategic and tactical nuclear weapon inventory of over 30,000 warheads.

THE TESTIMONY OF HIROSHIMA

The physical effects of nuclear weapon explosions were first seen in the cities of Japan a generation ago and have been carefully studied in the 1,100 test explosions carried out since World War II by the nuclear weapon powers—the U.S.A., U.S.S.R., United Kingdom, France, and China (and one by India). The United States has itself carried out over 600 tests. The Japanese suffered over 200,000 deaths from the two nuclear bombs of Hiroshima and Nagasaki in 1945. Since then a vast technical literature has been produced on nuclear weapon effects. But the matter is complex. So large an event as a nuclear explosion obviously has effects which depend upon many variables, and the consequences for people and structures at the scene or far away vary greatly with bomb design and with the circumstances of the event.

What a nuclear explosion does is to release a very large amount of energy with great suddenness (Figure 11). The explosive energy yield is not customarily expressed in standard energy units such as kilowatt-hours. It is rather described, in deference to the explosive nature and military history of the devices, by a direct comparison to the amount of conventional high explosive material—TNT—which would release the same amount of energy. The word "megaton" (MT) is used to indicate one million tons of "TNT equivalent," which also equals 1,000 kilotons (KT). Such a large measure was necessary once the weapons developers found the trick of thermonuclear energy release—the fusion of hydrogen rather than the fission of uranium—to make explosive yields almost unlimited. The use of the million prefix, *mega-*, is a reminder that the comparative unit—tons of TNT—is too small and therefore inappropriate, reflecting a historical comparison now obsolete.

Such large explosive yields as megatons are difficult to imagine, as the figures make clear. Another way to imagine nuclear yields is

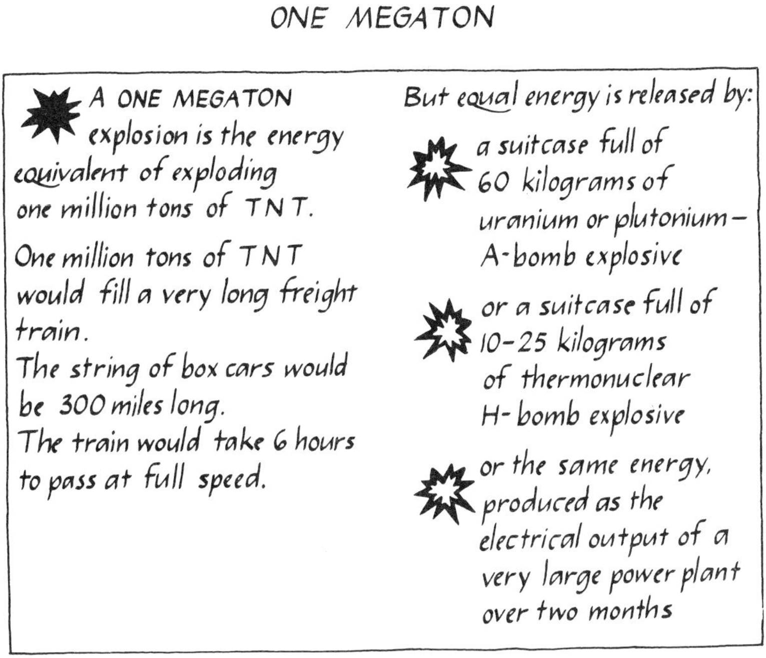

ONE MEGATON

A ONE MEGATON explosion is the energy equivalent of exploding one million tons of TNT.

One million tons of TNT would fill a very long freight train.
The string of box cars would be 300 miles long.
The train would take 6 hours to pass at full speed.

But equal energy is released by:

a suitcase full of 60 kilograms of uranium or plutonium — A-bomb explosive

or a suitcase full of 10-25 kilograms of thermonuclear H-bomb explosive

or the same energy, produced as the electrical output of a very large power plant over two months

FIG. 11

The megaton—metric jargon for one million tons (of TNT)—has come to be the unit of energy release in large explosions. That same amount of energy can be released in different ways over different times.

to compare them with conventional war explosives; the total explosive energy spent by the United States during the eight years of war in Vietnam was equivalent to about 6 megatons. The nuclear energy yield of weapons held by the United States strategic forces today, in comparison, can be estimated as some 7,000 megatons total.

The effects of nuclear weapons can better be understood by comparison with large natural disasters such as tidal waves or earthquakes. The earthquake energy, for example, which shook down the city of San Francisco 70 years ago can be estimated as some 10 to 12 megatons equivalent. A nuclear attack is best thought of today in terms of such great natural disasters. The inherent

A ROUGH GUIDE TO EXPLOSION YIELDS

CONVENTIONAL HIGH EXPLOSIVE:	
Standard hand grenade	½ lb. TNT equivalent
Usual time bomb left in public place	1 - 2 lb.
Field artillery shell	5 - 30 lb.
Typical big bomb, dropped from aircraft	¼ - ½ ton
Largest conventional bomb	10 tons

✸ 1000 tons of TNT equivalent = 1 kiloton (kT) ✸

NUCLEAR EXPLOSIVE:	
A small nuclear explosion	a few to 10 kT
Hiroshima - Nagasaki	10 - 20 kT
Modern MIRVd missiles / up to 14 warheads launched together: Poseidon C-3 in US subs	40 kT each warhead
3 warheads launched together: Minuteman III	150-200 kT each warhead

✸ 1000 kilotons = 1 megaton (MT) ✸

Single warhead in old standard ICBM: Polaris A-1 Minuteman I	1 - 2 MT
Older US land-based missiles: Titan II	5 - 10 MT
San Francisco earthquake energy, for comparison	10 - 15 MT
Largest known Soviet land-based missiles	20 - 25 MT
Very large bombs carried in aircraft	15 - 25 MT
Largest man-made explosion ever USSR in 1961	60 MT

DAMAGE EQUIVALENTS

8 explosions of 40kT each
are equal in area of damage
to 1 explosion of 1000 kT, or 1 MT

FIG. 13

Energy alone is not the only measure of an explosion. Very big explosions overdamage—even vaporize—the center of the explosion, wasting energy. Hence the area of serious damage for typical targets increases more slowly than in proportion to the energy release. This has given rise to a rule of thumb here illustrated: the damage area of smaller explosions is expressed in terms of the number of 1-megaton explosions which would produce the same nominal area of damage. As shown, eight 40-kiloton explosions are about equivalent to one megaton in damage area, although it requires twenty-five 40-kiloton explosions (25 × 40,000 = 1,000,000) to yield the energy of one megaton.

difference is that nuclear attacks would be carefully arranged to take place in close proximity to the works and the homes of human beings; they would occur, not once in a great while, as for big natural disasters, but rather hundreds or thousands of times within hours (Figure 14).

The similarities between nuclear explosions and natural catastrophes go farther than the amounts of energy released. The specific

FIG. 12

The list gives the energy yield of explosions of all kinds. Notice the gulf between conventional and nuclear explosives. The largest conventional explosions—ammunition ships blowing up and the like—amount to a few kilotons.

TYPICAL DAMAGE FROM A 40·KILOTON EXPLOSION

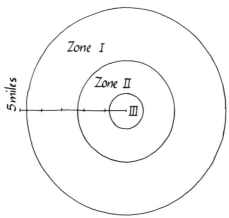

PHYSICAL DAMAGE	CASUALTIES
Out to 4.7 miles: General breakage of window glass, flying fragments, damage to weak structures.	Zone I, 55 square miles. Chance of death grades from about 5% at outer edge of this ring to 40% near its inner edge. Deaths caused by flying glass, fires, building failures.
Out to 2.4 miles: Severe burns to exposed skin.	
Out to 1.6 miles: Heavy structural damage to ordinary construction such as frame houses.	Zone II, 14 square miles. Chance of death grades with distance from 40 to 60%. Deaths caused by fires, flesh burns, collapse of roofs & walls.
Out to 1.4 miles: Ignition of clothing, light wooden pieces, etc.	
Out to .6 miles: Heavy damage to reinforced concrete structures: civic buildings.	Zone III, the innermost two square miles. 90% of persons present can be expected to meet death or grave injury through flying objects, collapse of whole structures, lethal radiation, thermal radiant burns, firestorm.
Out to .1 miles: Buried concrete & steel shelters and hardened missile silos made useless.	
Out to 200 feet: A crater of this radius is produced by an explosion at ground level.	

effects produced by a nuclear weapon also find similar counterparts in the violent events of nature. Thus, the *blast pressure* is similar to high hurricane winds which destroy structures and fling people and property about; the *intense radiant fiery heat* is like that of a volcanic eruption, which ignites fires in forest and city and burns exposed living flesh; the *ground shock* stimulates the effects of an earthquake; a weapon exploded in a river or harbor raises *enormous waves of water* to roll over the shares—waves similar to tidal waves. In addition to blast, heat, shock, and waves, nuclear weapons produce amounts of penetrating radiation unprecedented by any other phenomenon. This happens both at the time of the explosion, and later, in the form of fallout, whenever and wherever the radioactive by-products of the explosion settle to the ground. This radiation can be life-threatening to organisms of all kinds, even though it is usually unperceived by the senses. The effect has no equally acute counterpart on earth (though it has in exploding stars).

Estimates of all these terrifying phenomena can be made more or less reliably, given the complex conditions of each event. Wind and temperature, terrain, bomb design, height of burst, air transparency, heat-reflecting snow or absorbing rain, and other such variables all affect the result of a nuclear explosion. The target structures— material, human, and social—codetermine the effects as well on the actual targets. It is true that the higher the energy release (the greater the megatonnage), the more damage one can expect. But it is not true that these numerous effects all increase in simple proportion to the energy release. For most effects and most bomb designs, the damaged area increases with the energy release more slowly than in simple direct proportion.

The effect of this is that as a general rule, more damage can be done with several small weapons than with one large weapon. Thus, one

---

FIG. 14

The diagram maps the damage done by a 40-kiloton explosion at various distances from the point of impact. Plainly the variation is more or less gradual; the zone boundaries are rough classifications and refer to so-called prompt effects. Long-term effects, such as delayed cancers, would be in addition to the above calculations.

FIGS. 11 to 14

See notes on text.

might expect the damage of a 1-megaton weapon (one million tons of TNT) to equal that of twenty-five 40-kiloton weapons (altogether, also one million tons of TNT), but that is not the case. A 1-megaton weapon may produce an area of damage equivalent to only eight 40-kiloton weapons. Similarly, one 20-megaton weapon may do damage equivalent to only about seven 1-megaton weapons (Figure 13).

It is interesting—if a little academic—to apply this law of change over the whole range up from 1 ton of TNT. On that basis a 40-kiloton nuclear weapon is equivalent in area of damage (ignoring the indirect and delayed effects of radiation) to only about 1,200 conventional 1-ton high-explosive bombs. Of course this extrapolation is unreliable across so wide a range of effects; but nuclear yields can be usefully compared over a range of a factor of 1,000 or more for many purposes. From these data one can draw a few indicative results, shown in Figure 14.

Figure 14 should make two points quite clear. On the one hand, for ordinary civil targets, nuclear weapons are devastating. One modest warhead can cause heavy damage and catastrophic loss of life over many square miles, a sizable city district. On the other hand, a single warhead must come quite close if it is to destroy a buried, concrete- and steel-reinforced strong point, such as a missile silo. The bigger the bomb, of course, the bigger the area of damage. A 1-megaton bomb would increase the given damage areas seven- or eight-fold. It turns out that for "hard" targets, those which are specially protected or deep underground, a more accurate aim— which means a smaller miss-distance—is economical compared to a gross increase of explosive yield, and hence of weight of nuclear explosive. In other words, one has a better chance of destroying a hardened silo with a low-yield but highly accurate weapon than with a high-yield but inaccurate weapon.

With test data, and with the experience of the two Japanaese cities which were attacked with weapons of yield between 10 and 20 kilotons, it is possible to estimate the nature and number of fatalities and other casualties in a nuclear explosion as a function of the distance from the explosion. Of course, this estimated human destruction is contingent upon many variable circumstances, but it is no less likely to underestimate the results than to overestimate them.

Figure 14 shows potential damage relatively close to a nuclear explosion. There is also considerable threat to life at great distances from the burst. The fission product dust from an explosion drifts downwind, decaying as it drifts, and falling to the surface. Wind speed and direction determine this, as well as the height of burst, the nature of weapon, the nature of soil, and the weather. A typical elliptical downwind "plume" forms; people must either leave the area or find thickly shielded shelter before radiation exposure becomes fatal. Time and cover are crucially important. One may reckon that exposure to 300 or 400 rads (a unit of radiation exposure) will cause death in half the persons so exposed under the limited therapy available in disaster; the others will recover after some weeks of serious illness, though many among them may still develop malignant tumors many years later.

A wind speed of 15 miles per hour downwind from a 1-megaton nuclear explosion on the earth's surface yields a plume some 60 miles long by 10 miles wide; within this early plume, humans will be exposed to 500 rads within a few hours even behind the shielding provided by a brick residence. The plume lengthens, widens, and dilutes. It is more difficult to evade the radiation downwind, for it widens as it travels, but the farther it travels, the less dangerous the radiation. Evacuation would be prudent at least within an area 200 miles long by 30 miles wide within which a dose lethal to half the exposed people might be accumulated.

Many studies have been done on such a state of affairs; a conservative estimate predicts that the residents over 2,000 square miles have a substantial risk of radiation death or incapacitation. A similar area will be contaminated past safe use for a long time.

It seems probable that among the survivors who are lightly exposed many—perhaps one-tenth—will suffer radiation-induced tumors delayed over the decades after exposure. An estimate based on various exposures, including those at the Japanese cities, suggests that about one delayed malignant tumor can be expected for each 5,000 person-rads of exposure, with a similar number of genetic defects in the offspring.

A comparison is possible here with nuclear reactor accidents. The fission product release from an extremely serious accident in a large nuclear reactor would be 0.5 to 1.0 percent of the total fission product stored (3,000 kilograms). That is, after a meltdown and confinement loss, 15 to 30 kilograms of fission product would be released. This is equivalent to the fission yield of a 300- to 500-kiloton weapon. Thus, since weapons use fission and fusion combined, one may crudely take each 1-megaton weapon to produce about the same fallout as that expected from the worst of possible nuclear reactor accidents. (Over the first few days the weapon's fallout is much more dangerous than that from the reactor;

the reactor effects tend more to the long-lasting sort, which can be avoided more readily by evacuation.)

There are global risks as well involved with the use of nuclear weapons, although they are poorly known. There exists at present worldwide fallout from 200 megatons of fission yield produced by all past atmospheric nuclear tests, now much decayed. The full explosive nuclear power now held by the United States alone might raise that by some 25 times. If that megatonnage were to be used, one would expect about ten million induced tumors worldwide in this generation and several times that number of genetic defects passed on to the future. The ozone layer worldwide would be seriously affected by a many-thousand megaton strike; this would place crops, plankton, and forests at risk, a risk now believed repairable but, obviously, not known for certain to be so.

A number of populous nearby powers have a large stake in the matter of any future nuclear war, for the fallout is not likely to be spread with complete uniformity. Just as the countryside downwind from a military target or city will receive a very heavy plume of fallout, so China and Japan lie downwind from most of the planned impact sites for American weapons. They face serious danger, as does Atlantic Europe.

On the other hand, even the tens of millions of delayed tumors and genetic defects add only a fraction to those expected from natural causes among the world's four billion people. Although the safety factor is not as large as prudent public health officials might hope for—perhaps we are safe by a factor of only 100—it appears that all-out nuclear war with the present weapon stocks would not end human life worldwide as far as fallout effects are concerned. Whether some unforeseen subtle consequences for the climate or the ecology might turn out to be fatal is one more uncertainty of this grotesque situation.

TWO SCORPIONS IN A BOTTLE:
POLITICAL OUTCOME OF THESE WEAPONS

The logic of nuclear deterrence is still cogently expressed in the metaphor of two scorpions in a bottle, coined decades ago by nuclear physicist Robert Oppenheimer. True, times have changed. Several toxic centipedes now scurry around the two scorpions, and a number of other smaller creatures dream of acquiring stings. When only one scorpion had a sting, it thought of striking first, either in anger or to prevent the maturing of the other. But now a first strike would yield little gain to either, for the other can always strike back, at least so long as it possesses a sting which will survive the initial attack. This accounts for the rising importance of hidden weapons, like those under the seas, which are sure to survive a first strike. Given their secure, strongly toxic stings, the two scorpions are mutually deterred if they choose to survive. The ironic phrase employed to describe this standoff is "mutual assured destruction" (often abbreviated MAD): The population of each side is held hostage to the other, with no means of defense. The world remains in that terrifying state of deterrence today, as it has for a quarter of a century.

Nowadays there is a new idea about how nuclear weapons could be utilized, explicit in the U.S. Holding the great final stinger in reserve, one scorpion can yet punish the other by small, well-aimed stings in foreparts or legs, less than fatal, counting that the other will not risk mutual destruction by swinging the deadly tail. Might even enough of these pricks somehow weaken the great tail itself, before it is brought into final self-destructive use? Maybe. Such thinking has rationalized a great proliferation of small nuclear warheads, able to strike extremely close to any known target spot, and thus able to destroy military targets which are safe from more distant but larger explosions because they are of unusually strong construction or even buried deep underground. (In Chapter 7 we will discuss these points in more detail, without the metaphorical cover.)

We see no justification whatever for such profound risks. We will show that the deterrence of nuclear exchange—the posture of the two scorpions—is the only justifiable use for nuclear weapons; but this certainly depends on an air of reason and restraint inside the tensely balanced bottle. Our sense of timely action is to begin to reduce the venom of the stings while they remain deadly. The point is not at all who has the bigger sting, but that the stings are mutually fatal.

The tables and calculations of the previous pages should carry the grim argument plainly enough. Such measures of human and physical destruction are beyond precedent; they represent a misery and an injustice unique even in the terrible history of human conflict. We do not present these data lightly. The numbers can tell only a portion of the truth; the rest is a more human tale, a tale of families and crops and treasures of the world's culture exposed to apocalyptic terror, with one new Horseman called Radiation. No sane leader ought to expect gain from the use of nuclear weapons. It seems probable that the world's leadership, which has for almost 40 years carried us all up an escalator of potential damage, remains confident that it can still control unacceptable use of the weapons, and seeks to derive from their mere possession only the political benefits of prestige, power, veiled threat, and "superiority," without actual use. Our fear is plain: Such self-confidence in those with power is a defiance of history. Once it was called *hubris*; it remains a courting of the judgment of fate. For us who write, the only safe way is to wind the danger down.

7

WARFARE AT STRATEGIC RANGE: THE TRIAD

The accurate delivery of nuclear weapons to their targets—even allowing that the very largest weapons wreak havoc over many square miles—is a key part of warfare technology. Cannons, short-range aircraft, and missiles simply placed on the ground are possible means of delivery over modest distances, but across oceanic and intercontinental distances special means must be used. The system employed must be clearly reliable and adequately accurate under all conditions of weather and climate. More important still, the scorpion must be able to strike back even if it is struck first, even stung by a surprise poisoned dart aimed first of all at its own deterrent weapons. The search is for a relatively invulnerable force for retaliation in case of nuclear attack.

The sprawling flat airbases for manned aircraft—strategic bombers—can approach invulnerability only if some planes are organized in an expensive active rotation aloft, or kept for years in an alert state, only minutes away from flight. The airfields themselves are distinctly vulnerable, even to rather inaccurate attack. Arrival at the target by air is not sure any longer; missile defense against relatively slow high-flying jet bombers is by now highly effective, and the bombers themselves now must penetrate at low altitude and launch their bombs on short-range missiles to have any strong hope of striking their targets. Manned bombers might be recalled even after some hours in flight, which might offer certain advantages, but the retaliatory missiles could have already reached their targets.

Large ballistic missiles can carry nuclear warheads orbiting through space anywhere, anytime. Their launch tubes can approach near-invulnerability, providing only that they are hidden or shifting. Missiles in their silos on land—the ICBMs—are visible from orbit and hence targetable well in advance; they cannot be thought invulnerable. Only by launching the retaliatory missiles upon receipt of the half-hour warning notice given by detection of incoming hostile rockets can land-based ICBMs be sure of striking back. But this reaction risks elevating a mistaken warning of attack into worldwide holocaust within the hour. Various schemes are now proposed to fly missiles around in large aircraft, or place them on land in mobile launchers, shuttling them about in a "shell game" from one tunneled hardpoint to another.

There is a worldwide consensus that the depths of the sea are the securest hiding place for nuclear missiles. Four powers now keep missiles in submerged long-range submarines, scattered throughout the oceans (Figure 18). A number of such vessels, reliable enough so that some are always out on the high seas, shifting and hidden within the ocean depths, yet able at radioed command to carry out well-aimed strikes with deliberation from a great distance, represent the surest system of deterrence we know. This has been true for more than 15 years.

Unmanned long-range robot jet aircraft (descended by way of modern computers from the V-1 "buzz bomb" of World War II)— called "cruise missiles"—are now remarkably accurate. They are much cheaper and smaller than manned bombers. They lack, however, the ability of recall, and they are vulnerable to attack during their hours of low and relatively slow flight. As a result they do not seem to offer the U.S.A. an alternative to the swift ballistic missile in orbit above the atmosphere. They do, however, pose problems of international control for they can be hidden, are inexpensive, and hence might become numerous. The result of deploying long-range nuclear-armed cruise missiles—not yet in being—may be a new and critical race for expensive airborne defense systems: specialized aircraft with fast pursuit missiles. For, unlike defense against ballistic missiles, that against the slow air-breathing cruise missiles of long range looks practicable, if precarious. (Press reports already show signs of a Soviet system of look-down radars rising along their frontier; with fast short-range chase missiles, a well-directed force of fighters is likely to stop most cruise missiles short of interior targets. Less probably, such a defensive system lies ahead for the United States as well.)

Cruise missiles already have importance as short-range non-nuclear weapons against surface ships, and they might come to assume much of the role of manned aircraft for nonnuclear battlefields. Still, we certainly do not propose that the United States develop cruise missiles as strategic weapons. The use of long-range cruise missiles (as now the bombers use shorter-range missiles) to allow a strategic-bomber strike from 1,000 miles or more away has no significant deterrent function so long as we have numerous submarines and their SLBMSs.

TOWARD A SAFER STATE

The United States presently maintains a three-legged delivery system, called the "triad," for strategic nuclear weapons. (See Figure 15.) It numbers 1,054 ICBMs in silos on land plus 656 SLBMs on 41 submarines. There are also over 400 strategic jet bombers, in all more than 2,100 strategic launchers. The Soviet Union maintaihs about 2,400 strategic launchers, of the same general types, but at present armed with only one-third as many warheads as the United States; that gap is fast decreasing, however. Although ceilings have been placed by treaty on these incredible arsenals, the nuclear arms race continues and quickens, both in quantitative and qualitative terms. For example, the United States was deploying 50 new nuclear warheads a month a few years back, despite a public declaration of our interest to limit nuclear weaponry. We hasten to add that this was in no way a violation of the Strategic Arms Limitation Treaty, and more recently the U.S.S.R. has been legally building land- and sea-based missiles of its own at an even greater rate. (Therein lies the ineffectiveness of present treaties.)

How much foreseen retaliatory damage is needed to deter a war-bent leader? Evidently this is no objective military question, but rather a summing up of psychological intangibles. The horror of even one single targeted metropolis, burning at all its margins, flooded and smashed centrally, its countryside contaminated with lethal radiation, ought to deter any sensible leadership. Any American President would recognize that a single such event would cost in lives and in property more than any American war, even the Civil War, and it can quickly take place in hours. Moreover, it is plain that the existing explosive stores of both the superpowers can induce such a result, not merely in any one great city, but in a hundred cities all at once, to the point where intact centers of any importance would be rare. The bombs on Hiroshima and Nagasaki each took about 100,000 lives; today's nuclear arsenals threaten that damage with every single warhead out of the thousands that are ready.

War is no respecter of logic. Angry armies have fought on despite

US STRATEGIC FORCES

THE TRIAD

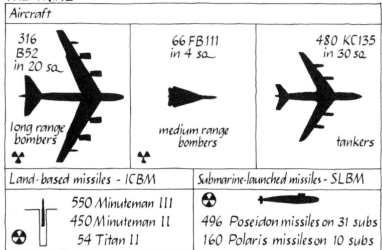

Aircraft		
316 B52 in 20 sq. long range bombers ☢	66 FB111 in 4 sq. medium range bombers ☢	480 KC135 in 30 sq. tankers

Land-based missiles - ICBM	Submarine-launched missiles - SLBM
550 Minuteman III 450 Minuteman II 54 Titan II ☢	☢ 496 Poseidon missiles on 31 subs 160 Polaris missiles on 10 subs

DEFENSIVE SYSTEMS

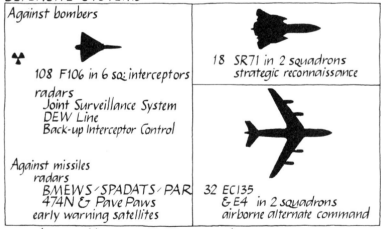

Against bombers

☢ 108 F106 in 6 sq. interceptors

radars
Joint Surveillance System
DEW Line
Back-up Interceptor Control

Against missiles
radars
BMEWS/SPADATS/PAR
474N & Pave Paws
early warning satellites

18 SR71 in 2 squadrons
strategic reconnaissance

32 EC135
& E4 in 2 squadrons
airborne alternate command

☢ nuclear capable weapon system   ☢ nuclear weapon system

FIG. 15

The active strategic forces of the United States are capable of delivering many nuclear weapons to any place on earth. Offensive nuclear strategic forces are terrible losses. Often they were unreasonably sanguine about escaping obvious destruction. Here we touch our deepest fears; the ultimate disaster of a large-scale nuclear war can best be precluded *at its source*, the overpowering quantity of weapons both superpowers hold. The probability of a rational decision to set off nuclear war is low; so much we have seen in 30 years. But the stakes are so high that we need to guard humanity against unforeseen traps in the maze of history. Some way must be found to reduce, slowly and prudently as may be, yet unmistakably, the catastrophic megatonnage of the present systems. In the long run, such capabilities can have no reasonable purpose, yet they pose an unending threat. If used on a large scale, their effects may well include epidemic tumors, depletion of the ozone layer, destruction of ocean plankton, even major alteration of climate mechanisms. No country could be wholly safe.

These threats present a steady worldwide risk greater than any from peacetime plutonium commerce, freon sprays, or supersonic transports. Neither a single accidental explosion nor the danger of a terriorist group with one homemade plutonium bomb or some radioactive dust is so terrible as are the regular military forces of the U.S.A. and U.S.S.R., with thousands of well-tested devices at the ready, each accurately aimed at some city, port, or underground post.

---

based on the triad, the term of the three modes of delivery, by land-based ballistic missiles, by undersea missiles, and by missiles or bombs carried on aircraft. The role of the tankers is to refuel bombers in flight, making possible heavy loads over the longest transoceanic ranges.

The defensive systems include active defense against bombers, but not against ballistic missiles, by the interceptor aircraft listed. These aircraft have available about 1,000 nuclear-weapon-tipped air-to-air rockets. For reconnaissance, the SR-71s are the fastest and highest-flying of all aircraft, some 2,300 miles per hour at 80,000 feet. The other planes listed maintain airborne control and command posts. The main radars and satellites are shown.

The strategic forces include also five submarine tenders, big ships which care for the subs in port. There are other radar and special warning systems; see DEFENSE SECRETARY FY 1978, pp. 142–3. The Reserves operate interceptor squadrons and many refueling tankers, MANPOWER FY 1978, IV-3, IV-5. B-52 is 158 feet long overall; and Minuteman is 60 feet long.

This cannot be regarded as a simple moral scruple. Indeed, the invocation of morality in this context of destruction must end in shame. We Americans are now targeting forests, cathedrals, and mothers, in unprecedented numbers, as hostages. The essential point is prudence; the powers must move steadily to a system of less danger. No physical argument fixes the minimum deterrent which the U.S.A. needs. The deterrent should begin to dwindle. Certainly we hold little brief for those present systems—like aircraft and ICBMs—which cannot survive attack.

We know of no serious counterarguments which can be based strictly on deterrence of large-scale nuclear exchange. (In the next chapter we will consider the important issues which arise from other perceptions of the utility of nuclear war for national policy.) Somewhere a start must be made.

OUR PROPOSALS

We therefore propose a series of sweeping changes in our strategic offensive forces. In brief, we recommend the steady elimination of nearly all land-based missiles, the elimination of the manned strategic bombers, but the retention of most of the submarine ballistic missile fleet, without the new Trident submarines now being built, and with a measure of reduction in the force loading of those subs we retain. We urge withdrawal of our sub bases to U.S. territory, and an end to growth in our present strong force of antisub attack submarines. Certain diplomatic proposals accompany the changes in military force. We proceed to discuss and to justify all these related actions in detail in the sections which follow.

1. The United States should gradually eliminate its land-based missiles, the ICBMs.

Step-by-step, publicly and clearly, the United States should retire the land-based missiles in their silos. A good rate might be three to five years of equal steps, say one-tenth or 100 missiles of the present force retired each year, beginning with the older Minuteman IIs. The rate might then be increased. If U.S. perceptions of the international scene then warrant, as we hope and expect, the reduction should continue until the whole land-based force is gone.

As a hedge against some quite unexpected new technical threat to the security of our sea-based nuclear forces, the United States might retain a couple of wings of the latest Minuteman III, say up to

100 hardened silos, beyond that time. The limited uses of so small a force are explained in Chapter 8.

2. *Eliminate U.S. intercontinental strategic bombers by stages.*

We recommend similarly that the force of B–52s and FB–111s be reduced, say 50 retired each year. The latest B–52 models, that is, the B–52Gs and Hs, a total of about 240 planes, might be retained for up to five years. But the eventual goal should be zero. Again that stance could be reached in six to eight years, if experience warrants.

These two measures in no way make the U.S. deterrent less credible. They would rather represent the belated rule of reason. It is not easy to examine the American "balanced strategic deterrent," with its missiles on air, sea, and land, without coming to feel that it has been the rivalry of the two more technical armed services, the U.S. Air Force and the U.S. Navy, as much as the security of the U.S.A. itself, which was primary in maintaining triad, the triple delivery system of ICBMs, SLBMs, and bombers. We do not need triad. It is not real defense, nor is it even a simple waste. It is an expensive liability, and it tends to provocation. Strategic cruise missiles are no different.

3. *Retain the submarine-launched ballistic missile system with reduced numbers.*

The United States should reduce the number of strategic missile submarines from the present 41 to 31 boats, the number of the more modern classes. This fleet, with each boat carrying 16 missiles (496 missiles in total) and each missile with some 10 to 14 warheads (Figure 16), is a formidable strike force—at least 4,960 nuclear weapons (Figure 20). Even allowing for boats off station for refit and for missiles off a target, a full salvo means 300 missiles or about 3,000 nuclear explosions. While submarines are in port for refit and repair they are not hidden, but surfaced and open to missile attack. At any given time more than half the U.S. missile submarines (55 to 60 percent) are expected to be safe at sea. In the U.S.S.R., by contrast, only about 10 percent of the boats are kept safe on station.

In the next chapter we enter more closely into the question of how

MIRVS

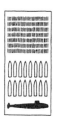

16 missiles are carried on one missile submarine. There are 10 to 14 warheads on each missile, each of which can be aimed at an individual target. One warhead on a Poseidon undersea missile yields 40 kilotons.

One launch silo buried in North Dakota or Montana holds one Minuteman III missile. Every such missile carries three warheads, each of which can be aimed at an individual target. One warhead on a Minuteman III missile yields 200 kilotons.

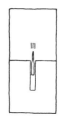

FIG. 16

What MIRV means: multiple independently-targeted reentry vehicles. (*Reentry* refers to coming back into the atmosphere after the flight in orbit through space.) The warheads are flung off from the vehicle one by one at the right time programmed to reach quite different points. Note how numerous are the targets open to even one submarine.

STRIKE & PLUME
THE AREA OF DAMAGE BY A SINGLE MISSILE

MASSACHUSETTS

Worcester.

Springfield.

Providence, RI.

0   25   50   75 miles

FIG. 17

The map gives an impression of a single strike by one missile with 14 warheads against Boston, a metropolitan area of about 3 million people. The damage circles are somewhat generalized, but conservatively sized for a 40-kiloton the missiles should be loaded—how many warheads they should carry; it is sufficient here to observe that a wide range of loads remains open to debate. The *minimum* load would be one Hiroshima-like warhead per *boat* (not per tube), still a forceful deterrent, capable of 20 Hiroshimas in a couple of hours. The *maximum* on station would be 300-plus tubes (16 tubes to each boat), each tube with 10 or more separately aimed warheads, enough to wreck every Soviet and allied city and industrial district, plus many extra military targets. (An intermediate load of *one* warhead per missile *tube* is discussed later.)

Since it is the number and quality of the moving, hidden submarines which fixes the invulnerability of the deterrent, we have chosen to retain the full force of up-to-date boats at its present size, even though we propose to reduce their weapons load. Such a force is so mobile and so scattered that no plausible improvement in antisubmarine detection and kill capability is given any chance of finding a large number of these submarines simultaneously as far ahead as one can now foresee.

4. Cancel Trident.

We oppose as unneeded the much bigger, still more expensive Trident submarines, the first couple of which are under construction. They are faster, quieter, and have a larger missile range than do the present standard Poseidon submarines, but this is not all clear advantage, for their larger size might make the new ships a little more detectable by some active form of sonar which sends out acoustic pulses and seeks the echo. Nor does it seem urgent to be able to fire immediately at a

---

warhead. (See also Figure 14.) Note the fall-out plume downwind; the wind blows most often as shown from the northwest quadrant. An explosion in a harbor can send huge "tidal" waves of radioactive water pouring over the city center. This terrible effect is not fully taken into account here. Even so, we reckon about 1.3 million people dead or seriously injured. There might be up to another hundred thousand delayed deaths along the plume. Consider what a couple of thousand of these strikes and plumes might do.

FIGS. 16 and 17

See the notes to the text.

longer range of 4,600 miles. The present U.S. Poseidon boats, with somewhat shorter-range missiles, could take the extra day or two to move up into firing range if they were roaming too far away when use was first indicated. (They are even now effective over nearly 3,000 miles distance.) We therefore also do not recommend producing the new longer-range missile, the C–4 Trident. The Poseidon missile, by 1977 deployed on all thirty-one of the newer U.S. boats, offers sufficiently swift and assured destruction.

5. Relocate sub bases.

We propose withdrawal of the missile submarine ports back to territory under the control of the United States. (Guam, Hawaii, and Charleston, S.C., are already in use.) This means that submarine withdrawal from the fleet base in Holy Loch, Scotland, should at once be discussed with the British. (Rota, our present submarine base in Spain, will be vacated by treaty in 1979.) It is better for us to have full independence in submarine basing.

ARE UNDERSEA MISSILES SECURE?

A short account is here appropriate of the nature of these submarines, and of possible threats to their survival. The strategic missile submarines are nuclear submarines in two distinct meanings of the term. So far we have spoken only of their weapons, explosive thermonuclear warheads. But they also have controlled nuclear power reactors, which power the engines to drive them through the seas at unprecedented speeds underwater, at 30 to 35 miles per hour. The submarines also have enormous endurance, free from any need for air or for refueling, during voyages long enough to go around the world and more. They can and do remain submerged and in constant motion for two months at a time. Able to receive radio or acoustic orders while submerged, they need not break the surface at all. They navigate mainly by inertial means, though sometimes they use clever schemes for an occasional calibrating peek at satellites or stars.

Only one predator has the potential to threaten them at all: the nuclear-powered antisubmarine (or "hunter-killer"). This nuclear-driven cousin of the missile boats carries no long-range missiles, but because of its reactor-driven engines it matches the great speed, range, and long underwater endurance of the missile subs. There is no way yet, in spite of decades of costly effort, to detect and localize a submerged submarine reliably, once it is lost in the echoing vastness of the open sea. But a nuclear hunter-killer submarine has a chance to pick up and shadow a missile submarine as the latter leaves its home port. Of course, once in a while, by chance or as a result of various countermeasures, the contact will be irretrievably broken. But given a few hunter-killer submarines assigned to each missile boat, the scheme might hope to destroy each missile boat as it tries to fire. Such taut shadowing has been carried out experimentally from time to time by the U.S. hunter-killer subs; were it a steady practice by either the Russians or the Americans, it would raise grave threats of conflict. Other surveillance schemes are possible, and an effort to break the distant command links with the subs might be tried. Still, deterrent safety is found in a number of missile boats sufficiently large that the opposing side's hunter-killer submarines do not considerably outnumber them. For these reason, we recommend retaining a good many missile boats (while offloading a portion of their nuclear explosive, the "overkill"). Extensive shadowing is then in no way feasible. We would also avoid building and deploying so many U.S. nuclear-powered antisubmarines that the Soviet Union would see our force as a direct shadowing threat to its undersea deterrent. For the present, the U.S.A. has a number of antisub submarines dangerously close to sufficient to allow shadowing Soviet missile submarines at sea.

All this deadly game of tag underseas seems so incalculable that most students discount it as any realistic danger to the undersea deterrent. One point is sure: However real it is, at present this is a one-sided danger to the Soviet ballistic missile subs, with very little threat to our own. For the U.S. has many powerful hunter-killer subs to send against the small Soviet missile fleet now kept at sea, while the U.S.S.R. has few—if any—capable hunter-killer boats to threaten our large, speedy, and quiet strategic fleet. Recall that the diesel submarines of yesteryear—very many are still listed in the Soviet Navy—are not important in this calculus, for they lack by far adequate endurance and speed.

The Soviet submarine-based deterrent is not yet so clear a success as is the American. The United States tries to keep its boats to a firm schedule: This is normally two months at sea submerged, and a third month spent in base for refit, with two complete crews taking alternating tours of duty. The planned longer-time repairs reduce the fraction of boats on station from the planned 60 to 65 percent to something over 50 percent, still an enviable record of skill and devotion. The Russians, without regular double crews, and with a slower repair cycle, presently manage to keep only *11 percent* of their first-line boats on station. They thus have at sea six or eight such boats, many fewer than U.S. hunter-killer submarines are capable of shadowing. This may help explain the number, variety, and payload of their new land-based missiles (Figure 18).

OVERKILL: WARHEAD NUMBERS AND YIELDS

The quantity of deterrence is a subtle topic. Given assurance of nuclear weapons striking their targets, how much explosive does one have to detonate?—indeed, how much can one bear to use? The present full amounts are a real environmental hazard; although public studies are not precise, multiple nuclear explosions imply a variety of catastrophes even to populations far from the aiming points, even in other countries. The dusty eruption of one volcano—Krakatoa—changed world climate a little for a few years; our nuclear weapons offer simultaneous Krakatoas by the hundreds. On the other hand, even a few dozen warheads of Hiroshima size would mean unprecedented punishment for an enemy civilian population. It is possible to take some historical tragedy, say the German invasion of the U.S.S.R. in 1941–43, and to undertake arbitrarily to double it suddenly in deaths and destruction and the denial of use of land; this would represent 20 to 30 million dead and the loss of an area the size of the United States east of the Mississippi. That could be managed with a payload much smaller than that now in place. Public debate should arise on the morality and wisdom of so great a threat of destruction. We feel sure that Americans can find safety with much less, insofar as safety can at all arise from deterrence.

What then is the significance in damage of the present nuclear weapons inventories? Since we recommend against continued maintenance of large forces of land-based and airborne strategic missiles and bombs, we can restrict our discussion here to the Poseidon missiles now installed on the 31 American missile submarines of the *Lafayette* class and later, the largest and fastest U.S. submarines now in service. (We propose to retire the 10 older

WORLD STRATEGIC NUCLEAR FORCES COMPARED

		Maximum number of targets	Damage equivalent in megatons	Weight of payload in tons
USA	land-based	2154	1460	1100
	undersea	5120	830	600
	airborne	over 4000	4400	11000
	TOTAL	over 11000	6700	over 12500
USSR	land-based	2650	2950	3900
	undersea	910	860	650
	airborne	270	780	2400
	TOTAL	3900	4600	7000
UK	undersea only	64	70	100
France	land-based	18	5	6
	undersea	64	45	40
	airborne	32	few	250
	TOTAL	120	50	300
China	land-based	50-80	20-60	30
	airborne	80	200	250
	TOTAL	150	250	300

FIG. 18

This figure gives a comparison of nuclear strategic forces. Three points are central: first, the various ways in which the forces can be compared; second, the diverse nature of delivery (the triad); the third, the big difference between the superpowers and the rest. It is to be remarked that most of the delivery vehicles of the U.K., France, and China are not of full intercontinental range; on the other hand, they need not be, given their function and the geography. (See also Figure 4.)

MILITARY BALANCE 1977–78, national entries, and Table 1,A,iii, somewhat modified by the authors. JOINT CHIEFS FY 1978, pp. 22 and 31.

Polaris submarines.) These 31 boats mount sixteen launch tubes each; each tube holds a Poseidon missile that may carry up to 14 warheads, each independently targetable within broad limits. The yield per warhead is between 30 to 50 kilotons equivalent of TNT,

THE CONSEQUENCES OF A NUCLEAR STRIKE

On Soviet cities by US missile submarines:

Size of attack in number of submarines	40 kiloton Warheads	Damage equivalent in megatons	Destruction of industry	Persons dead
5	800	100	59%	37 million
20	3200	400	76%	74 million

On US cities by US missile submarines:

Number of submarines	40-kiloton warheads	Damage equivalent in megatons	Dead or seriously injured: Target areas only	Dead or seriously injured: Total, including fall-out plumes
1	160	20	16 million	about 18 million
2	320	40	32 million	35-40
4	600	75	48 million	55-60
9	1200	150	60 million	70-90

FIG. 19

These figures report the grim symmetry of nuclear strikes against cities and other populated areas.

The first (put into the record officially a decade ago by Secretary of Defense Robert McNamara) estimates what is to be expected from strikes of varying yield on the Soviet Union.

The second describes the effects to be foreseen after an attack on the cities of this country by missiles like those of U.S. subs. These were computed by the authors in a simplified way to stand as a rough check of the more elaborate official estimates, which used Soviet area and population statistics, less available to us than are those of the 1970 United States Census. We were also then able to extend the range of strike size considered. (See the notes for details of our calculations.)

We have included some estimates of the effects in the plumes in addition to compared to the nominal Hiroshima yield of 14 to 20 kilotons. We can conservatively assign to each warhead an area of major damage of about 15 square miles, within which most housing will collapse, most strong civil structures will receive permanent damage, and most persons not deep underground will be promptly killed, without taking any account of delayed radioactive effects, nor the severe fires. Much secondary damage, of course, extends beyond this boundary. (See Figure 17.)

Considering the urban and industrial areas of the U.S.S.R. (the U.S.A. may be somewhat more urban, though Soviet industry is probably the more concentrated), authoritative studies in the early 1970s by the Department of Defense led to the rough estimates of the immediate effects of nuclear strike shown in Figure 19.

There is no doubt that the present nuclear strike force, with its 8,500 to 9,000 warheads, is greater than any clear need for retaliatory "assured destruction," even allowing for many targets within other countries, boats off station, warheads which fail, the effects of defense shelters and civilian evacuation, and so on. Since the number of warheads judged by some experts to be enough to destroy all the significant unprotected above-ground civilian targets in the Soviet Union and China is under 1,000, the rest represents strikes against a wide variety of military and industrial targets, many of them especially protected, like submarines in harbor and missile silos. The U.S.S.R. has a similar overkill capacity, chiefly land-based at the moment.

The calamitous submarine weapon load can and should itself be reduced, primarily a matter of internal warhead changes, or of off-loading warheads. But how far down the United States can go and how fast is not clear; it requires public thought and debate as to what constitutes an adequate deterrent. It seems to us that a much more stable world—because a more reasoning one—will be at hand once the total U.S. nuclear strike force is reduced from its present level.

---

those in the target areas. Neither estimate takes realistic account of the delayed social effects, life-threatening in the extreme. How will the survivors feed and warm themselves, with most transportation networks smashed and most stored goods lost? Even the next crops are endangered by the effects of fire and of radiation. How will the distribution of scarce necessities work in this time of chaos and destruction? Such uncertainties are in addition to the disastrous estimates given in the figures.

A FORMIDABLE DETERRENT

$31 \times 16 \times 14 = 6944$

$100 \times 3 = 300$

Warheads $= 7244$

Do we need to wreak sudden utter destruction on a whole people, and to threaten the world environment? Or is it enough to kill an important fraction of the nation, and a major part of their industrial wealth? It should be possible to reduce the 5,000 to 7,000 ready 40-kiloton warheads now on our recommended 31 boats down to one warhead per missile, 16 tubes per boat, for a total of 496 warheads. This is an adequate deterrent, intermediate in sizing; allowing for boats not on station and warheads not on target it would still guarantee about 40 equivalent megatons delivered—a damaged area like 600 Hiroshimas—and would destroy more than a third of Soviet industry at once, with the probable prompt death of 15 to 20 million people. We say nothing of the raging fires, the contaminated lands, the burned and injured, the epidemic of tumors, the dearth of food and fuel and shelter in the winter to come, the scattered fabric of the nation. The sure retaliatory strike would come, whatever they might do, within hours or days at most after an American decision to strike. Ample deterrent?

It is not clear that a reduction in U.S. warheads would be fully credited by the rest of the world. It is an internal change not easily verified. Nevertheless, we recommend proceeding with reduction in the submarine payloads, slowly and openly. This will not risk our assured deterrent, and may begin the necessary process toward a world of some safety for us all.

Beyond this, the whole long meaning of deterrence needs study. Are there any ways to hold the leaders hostage more than the populace? Can the United States be more cautious about when, how, and against whom it retaliates? There is room here for much strategic study, in the spirit, not of increasing overkill, but of

---

FIG. 20

The maximum nuclear deterrent of our proposal: thirty-one subs, sixteen missiles per sub with up to fourteen warheads per missile, plus one hundred silos with one missile per silo, three warheads per missile. The text makes clear why we are hopeful of reducing the loading of warheads on the submarine missiles considerably, even though it has no easy verification or budgetary effect. Compare the 7,000-plus warheads here with the calculation in Figure 19. Indeed, a formidable deterrent!

FIGS. 19 and 20

See text notes to close of Chapter 7.

mitigating the awful disaster which the United States' best present defense—deterrence—so far threatens the world at large.

It is to us—and we expect to our readers—likely that the growing problems of nuclear proliferation among other smaller nations will not be solved in a world where we Americans and the few biggest powers flaunt enormous nuclear weaponry; the necessary precondition to that solution is probably the reduction of our own nuclear arsenal. Once the United States has achieved a level adequate for deterrence but no higher than necessary, we recommend further reductions by safe, gradual, and mutual steps, until one day such weapons are entirely gone, here and abroad.

THE STRATEGIC DEFENSIVE FORCES

Naturally, there has been an effort to *defend* the United States against strategic nuclear attack—not merely to *deter* the attack. Many will remember the hopes for a supposedly effective antiballistic missile defense system, ABM. It was never completed. In 1976, the Congress acted to terminate operation of the single residual ABM site which the United States then maintained more or less as a pro forma equivalent of the system around Moscow. At best, these ABM systems would be easily overloaded by the dispatch of many missiles against them; the system in the U.S.S.R. plays a minor and dwindling role. Aviation circles recently surfaced a wild extrapolation of Soviet ABM capabilities to suggest a new breakthrough: a particle-beam missile defense. It would by no means be without natural and effective countermeasures, were it in fact realizable.

A defense against bomber aircraft is far more feasible, though by the same token the airborne threat is nowadays much less likely. (Note, however, that our bombers still hold half the U.S. equivalent megatonnage.) The U.S. Air Force operates an alert screen of supersonic all-weather interceptors with air-to-air missiles and cannon (the F-106 Delta Dart, first deployed around 1960) around the periphery of the 48 states. This force is decreasing in strength as the likelihood of penetrating bombers lessens in this missile age; it will in 1978 include some 108 aircraft (six squadrons) of this type, plus another half-dozen back-up Air National Guard squadrons, and a few forward fighter-bomber squadrons in Iceland and Alaska. These aircraft are directed by a radar and control system which is quite advanced (JSS), replacing the older centers of the so-called

SAGE system. This new ground Joint Surveillance System is shared with the civil aviation authorities of the U.S.A. and Canada. Airborne back-up radar and control (BUIC) are also provided. Deployment of the new interceptor proposed to replace Delta Dart in the 1980s can well be deferred, we believe; the present force represents no policy risk. Its scale and costs should continue to decline slowly. (See Figure 15.)

The U.S.S.R. has long faced a much more powerful intrusive bomber threat from our Strategic Air Command; Soviet interceptors are very numerous indeed, though the bulk of them are mostly of short range, suited to their internal defense task. Their control and radar net is also most extensive; it now lacks effectiveness mainly against low-altitude high-speed entry—such as by cruise missiles— according to the American Joint Chiefs of Staff. (This deficiency is slowly being repaired by new systems.) Right now the U.S.S.R. has a large inventory of nonnuclear surface-to-air missiles—10,000 to 12,000—which provide an effective defense against high-altitude subsonic bombers.

The United States no longer maintains homeland antiaircraft missiles for surface-to-air defense (except in Alaska and Florida). For warning of ballistic missiles and space vehicles we operate a series of radar and optical surveillance systems of high performance (SPADATS, etc.). The three Ballistic Missile Early Warning sites (Britain, Greenland, Alaska) remain in use and are complemented by special coastal radars against submarine-launched missiles (474, becoming Pave Paws) and by satellite-borne systems which altogether cover "all relevant strategic missile launch areas with at least two different types of warning sensors," according to the Department of Defense. One big ABM radar in North Dakota remains in operation.

Since not much more than 30 minutes will elapse between the launch of any ballistic missile and its impact, a 15-minute warning gains very little for the civil population or even the leaders. Its main points would be to enable the dispatch of the alert bomber force, leaving the vulnerable hangars on the wide airfields, and to allow the option of firing some missiles from silos. The submarines can react in hours, or even after days, with safety. These quick-warning systems are thus not of great use without the dyad or triad. To be sure, the direction and number of incoming missiles can be of prime importance for many decisions, possibly taken more wisely after a moderate interval. Still we do not recommend phasing out these somewhat redundant systems for the present time.

8

QUEST FOR NUCLEAR ADVANTAGE

A U.S. FIRST STRIKE?

For three decades now, a clear line has been set between nuclear and conventional warfare. Any blurring of that sharp distinction is a serious danger to the security of the United States in particular, and to the rest of the world as well. These weapons, with their unprecedented effects upon human society and the terrestrial environment, have no place in the execution of a wise security policy, save the single one of deterring nuclear war by the threat of retaliation in kind. That was the topic of Chapters 6 and 7.

We are well aware that no expressed U.S. policy, past or present, has given so restricted a place to nuclear weapons. While the United States held a nuclear-monopoly in nuclear weapons, and even sometime beyond, John Foster Dulles, President Eisenhower's Secretary of State and chief foreign policy architect, spoke of "massive retaliation at places and by means of our own choosing." It was clear what he meant: The United States might initiate strategic *nuclear* war should another power wage even a conventional war against U.S. central interests. In Europe the United States has led the massive deployment of what are called tactical nuclear weapons, with a clear indication it might well employ them first in the event conventional war threatened a NATO defeat. (Chapter 12 is devoted to discussion of the special problems of the tactical nuclear weapons.)

Once Americans were confronted at last by an adversary with a massive nuclear retaliatory capability against the U.S.A., they began to emphasize the importance of mutual deterrence. U.S. policy remains close to that stance today, though with major ambiguities.

But the effort to exploit nuclear weapons has not ended. On the contrary, certainly since 1974 under then Defense Secretary James Schlesinger, the United States has publicly planned more than a sure deterrent to nuclear attack. U.S. policy has sought to exploit American expertise in nuclear weapons and in their distant delivery for limited warfare, perhaps even over secondary issues. The Department of Defense certainly considers the intricate possibilities of an escalating nuclear attack of its own after any possible transgression across the nuclear line by others. The Department of Defense contemplates nuclear first use, in a number of opening games of strategic threat, expecting U.S. leadership in accuracy and control to offer some advantage. The doctrine seeks to turn nuclear war as a whole into a continuum, which U.S. leaders would have the option to enter at some small scale. That policy claims the possibility of a temporary resettlement of some U.S. city populations in mines and in improvised rural shelters, and the use of our nuclear missiles of high accuracy to destroy particular military-industrial targets within the U.S.S.R. in a carefully planned sequence.

This dangerous strategy has been called by various names. The most reassuring name is the offering of options for "flexible response."

We pay here explicit attention to recent arguments put forward by the Department of Defense and its friends who look forward indeed to constant renewal and improvement of the triad within the broad SALT limits. The argument has been clearly put forth officially by former U.S. Secretary of Defense Donald H. Rumsfeld. He had described the need for "options" other than the assurance of a strong second strike, options which allow a range of "responses short of full-scale retaliation." We hold that such options are both destabilizing and infeasible.

The U.S. Department of Defense has looked far beyond the deaths of tens of millions of people to the subsequent historical stages: "An important objective . . . should be to retard significantly the ability of the U.S.S.R. to recover from the exchange and regain the status of a 20th-century military and industrial power more rapidly than the United States." This goal is virtually beyond comment: In brief, we condemn it. It is a modern return to the policy of Tamerlane who threatened to plow the fields with salt. Just the same intentions are imputed to the U.S.S.R., with a show of evidence.

Secretary Rumsfeld states: "However much one might wish otherwise, popular and even some governmental perceptions of the strategic nuclear balance tend to be influenced less by detailed analysis than by . . . static indicators . . . launchers, warheads, mega-tonnage, accuracy and the like." Here a reasoned conception of deterrence has given way explicitly in the Secretary's view to the perception of power, to image. As the report explained: "Accordingly, U.S. plans and programs and future U.S. offensive capabilities must be geared to those of the U.S.S.R." Those we can assume will then as now be geared, in the presence of a pretty likely *qualitative* inferiority on the part of the U.S.S.R., to provide more explosive power than has the U.S. This is the classical story of instability, feedback, and runaway.

But there is a path of less risk. To understand it, we outline in brief the technical history of the missiles over the last decade or so.

MIRV AND MaRV

A ballistic missile, as the name suggests, is rather like a gun of long range, in some cases worldwide range. The rocket fuel burns in orbit to power the upward motion of the warhead as though it were within a gun barrel hundreds of miles long; after burnout, the warhead flies freely through space and down back into the atmosphere and to the ground target like a bullet, with no further power or control. But in the late 1960s, a novelty appeared first in the U.S.A.

The new technique originally was designated MRV (*M*ultiple *R*eentry *V*ehicles). The "bullet" was made multiple, several warheads falling from one reentering rocket body in a fixed shotgunlike pattern or "footprint." Then around 1970 the United States began to deploy a still newer sort of missile: MIRV. (The letters stand for *M*ultiple *I*ndependently-targeted *R*eentry *V*ehicle.) Something quite new had now been added. The device was now given to the power to send its warheads to distinct but prechosen spots, in a pattern which could be programmed within limits to reach individual ground targets hundreds of miles from the initial aiming point.

This is MIRV (Figure 16). By 1978 all 31 U.S. Poseidon submarines will be fully MIRVed, with 10 to 14 warheads per missile, each for its own target. Our land-based missiles are now also in large part MIRVed, the mix of Minuteman missiles now contains 450 older missiles with single warheads, and 550 MIRVed, each with three warheads. (MIRV is much cheaper per target than one missile with one warhead in a silo.)

It is striking to recall that one of the original justifications for MIRV was the desire to overwhelm by sheer number and by decoy techniques the Soviet antiballistic missile system, now 64 antimissile launchers, never very effective and no longer even relevant against our thousands of warheads. (The U.S. deploys no ABMs.) MIRV instead now affords scope for the "options" of elaborate partial-attack schemes, with more warheads than plausible targets, and a tradeoff which destroys many enemy MIRVed warheads, sitting in one missile, for each one of our own that comes in first.

Now the United States is about to have MaRV (the *Ma*neuverable *R*eentry *V*ehicle) with two systems presently under development. Here the warhead can sense its own position by one or another means—stars, signals from satellites, even its own radar. It no longer flies like a bullet to its target but maneuvers in its final approach, correcting errors, the effects of wind and so on, with a great increase in accuracy of the position of the explosion. Such accuracy increases offer in fact a very much increased chance of destroying missiles in hardened emplacements even with the use of smaller yields. Yet MaRV is also acclaimed as a hedge against antimissile defense. The new Navy MaRV development, the Mk–500 reentry vehicle, is still named Evader.

Meanwhile, the newly ready U.S. Air Force Mk 12–A warhead and the NS–20 guidance system, now being deployed, will result in both increased yield and improved accuracy for the present Minuteman III MIRVs. Although this is not a MaRV, the Mk 12–A is intended to attack hardened targets. That of course implies first strike, or a step-by-step exchange. It is no use to fire upon empty silos. Developments in accuracy are of use mainly against underground "hardened" military targets, in what is called counterforce—that is, against the enemy strategic forces; they offer in our view a dangerous new incentive to the missile race, particularly since the U.S.S.R. must depend so heavily on its land missiles, given the U.S. attack submarine threat to the small Soviet sub forces now at sea. In fact, the sum of the new technical developments and the new "options" strategy might be judged not only counterforce but even counterdeterrent—its eventual grand aim to be able to destroy the entire Soviet nuclear capability in one single blow.

The cost of steady technical advances in the plan for 1978 amounts to $0.5 to $0.7 billion. We would cut most of them off now. Enough!

THE NEW TRIAD AND MISSILE X

Besides a new bomber, B-1 (moribund though not yet dead), and a new missile sub, Trident, yet another new strategic weapon is in active development—the Air Force Missile X (MX). (The Services thus seek a whole new triad.) B-1, whose major production was canceled in July 1977, is a supersonic, long-range, low-altitude penetrating bomber. Trident is a system of larger, faster nuclear missile subs with longer-range missiles. Here are costly, higher-performance updates of the present, of the B-52 planes, and Poseidon missile subs. They do not bring conceptually new strategic problems; they are unnecessary, and we recommend against them in Chapter 7.

But MX is otherwise: It is a new effort to make land-based missiles much less vulnerable. MX is intended to supplant some or all of the Minuteman IIIs, in a new way, from a shifting base—the "mobile mode" alluded to in Chapter 7. MX might shuttle by truck or train among shelters prepared for it: The point is to shuttle the missiles daily among so many possible sites that an enemy cannot hit them all. Missile X has other advanced properties: higher yield, increased number of independent warheads (including maneuverable reentry possibilities), and accuracy improved to a miss distance of 300 feet or less. In short, it is a weapon to kill protected targets, perhaps becoming a credible threat to the big Soviet land-based missile force, while their submerged force remains small. (Its use might of course only trigger a Soviet launch-on-warning!)

By Defense Department plans MX would first have been deployed in the very same fixed silos which Minuteman now occupies, but early in 1977 Congress objected. No firm decision has yet been reached on how the mobile mode will eventually be carried out. The cost of MX is not small; the estimates run at some $30 to $50 billion for a deployment of 200 to 300 missiles.

The objectives of Missile X are not only a lessened ICBM vulnerability, but also an improved hard-target kill option, part of the quest for advantage. It is perceived by the Soviets in just that light. We hold that it represents a provocative, unneeded, and dangerous step in strategic escalation. We conclude that Missile X should not be deployed at all, and its development stopped.

THE PRESENT DANGER

It is necessary to consider these new options in the context of the Soviet missile force today. All of us have read scenarios of foreboding, based mainly on the U.S.S.R.'s genuinely large and growing force of land-based missiles, and secondarily on the more conjectural claim of successful Soviet preparations for passive civil defense against nuclear war. Whatever the Soviets may intend by their powerful force, however they have been led to construct it, it is for us here to examine whether it might be rationally used to attack or even to overawe the United States. In this book especially we need to face the question. Our recommendation is to phase out most or all of the U.S. land-based missiles and bombers and so to come in a few years to base our deterrent force mainly on a "monad," the undersea missiles, approaching that state by gradual stages from the present triad.

First, we again outline the relevant differences among the triad launchers, but now in the narrower context of a limited attack.

1. The missile subs are hidden beyond quick attack. But they are both psychologically and physically beyond easy command reach; targeting details are harder to send to them, and confirmation of receipt is difficult and risky. Each of them is an expensive basket of many missile eggs; they are assets not lightly to be risked by the disclosure which sending off one single missile might make to the infrared satellite far overhead, or radar across the horizon. (Planners of the worst-case school assume that a sub is fatally compromised by launching even one missile.) The undersea missiles are not now accurate enough to assure the destruction of one strongly hardened target, like a buried silo, even with several of their warheads. (Their spread is estimated as around 1,500 feet from the point of aim.) They remain deadly, not only against cities, but also against factories, harbors, barracks, and other lightly protected valuable targets above ground.

2. The land-based missiles are clear targets in their visible and stationary silos, well-mapped from orbit. Though strengthened by much steel and concrete, and buried deep in the ground, they remain vulnerable to accurate attack. Thus they can be used first, or as part of a step-by-step exchange, or perhaps during the short interval of flight after the other side has launched missiles against them. But their aim is very accurate (the strike within 30 feet of the aiming point can be expected), and they can be dispatched one by one on

command, in confidence that complex new orders will be followed quickly and accurately. There is good two-way communication between the leaders and the silo fields; orders can be easily confirmed.

3. The bombers are slow to strike—hours of flight to approach their firing points—so they might allow a negotiated recall. The long subsonic path nearly to the target leaves them seriously vulnerable. Their dispatch is likely to trigger missile response long before their arrival. They weigh for little in the current discussion, though they might be sent against certain targets.

How can power advantage be derived from a big land-based force, without drawing the deterrent retaliation? Now, consider the gross scale of attack. If it is a matter of interest less than absolutely vital, one might expect an exchange of warheads strictly limited in number, say tens or even a hundred or two. Even with care taken to avoid population centers, such an attack would bring in a few days the amount of civil casualties on a scale comparable to any of our past wars, hundreds of thousands dead or seriously injured. Such an attack would justify retaliation of its own kind, a counterattack against some hundred targets in response. At the other extreme, an attack of many thousands or tens of thousands of warheads would be so severe as to wash out distinctions between the damage and that of a deliberate city-busting salvo. The intermediate ground is the "realistic" one, an attack which might be made against our 1,000 land-based silos, with loss of life limited to millions, instead of the many tens of millions of the all-out attack on cities. That strike would then, so we are told, be followed by some national blackmail enforced by the threat of use of the rest of the attacker's more numerous land-based forces, a couple of thousand warheads more.

Just what targets could be chosen by the attacker? The official Rumsfeld statement quite correctly, we believe, saw the first danger: that to our allies. No matter how close an alliance, no party will prefer the explosions in his own territory. So the less provocative attack is on American interests in Europe. Next most plausible, perhaps, is a creeping attack on some subs, on our overseas military forces, and even on silos in the U.S. Third, least likely but most scary, is a big attack of some 2,000 explosions against all our silos, bomber bases, and subs in port. Each of these is imagined followed by a blackmail threat, under the high penalty of use of the still unfired surplus of missiles against our cities.

The scenario demands reply. We need not consider all the

methods which might deter such an action. Its probability is low, its dangers many; the misuse of our own preparations to hatch schemes of our own for such risky preemption and blackmail is by no means out of the question. All we need propose is a single deterrent of this course of action which will work in the circumstances. We submit that our powerful submarine force presently provides plenty of retaliation in kind. Three to five subs—up to one-sixth of our ready second-strike force—can threaten retaliation to some 1,500 to 2,000 soft-skinned targets, short of the cities. There are plenty of harbors, airbases, refineries, and barracks to aim at. Our last-ditch hostage deterrent remains as terrible as it was. The subs deter blackmail as surely as they deter a total attack. To fend off a limited attack you do not have either to kill the assailant, or knock the stick out of his hand; it is enough to be able to kick him really hard. The presumed attack would then lead to an *exchange* of some thousands of warheads, terrible military losses, and no winner to be found. The targetable land-based missiles are not relevant; let them go. True, the commands would travel slower undersea, and the leaders would be less confident of compliance, though radar and satellite would soon tell when the mission had been accomplished. (See Figure 20.)

These important details might be addressed by the experts; we judge the reliability of the redundant communications link is high enough now, and it could be increased at modest effort. The safety of our scheme is enormously better for all parties. The implicit match of silo for silo some propose is curiously direct; there may be goals not plainly expressed, from service loyalty and national prestige at the best, to hidden hopes for a blackmail venture of our own at the worst.

Even accepting the worst-case imaginations of the alarmists does not modify the logic of deterrence today. The silos are obsolescent. The subs have plenty of power, safety, and accuracy to deter a heavy attack on U.S. military and economic targets by their ability to react in kind. (The unlikely brief mass evacuation of cities to reduce losses does not affect the issue. For the cities, their food, stocks, fuel, factories, and treasures are all still there under the gun.)

We need only take care to maintain communications with the subs by redundant means. The present force loading of the subs is more than ample. We have leeway there to cut back, if not by the factor 20 implied in Chapter 6 as a goal, at least by a factor of four or five. (On this grave question of loading, an official study is needed at a high technical and political level, not by alliances between inexpert officials and experts from the aerospace industry!)

This discussion comes to a central dilemma. The leaders and their military counselors do not easily accept the hard truth of the present world: It is risky to seek even subsidiary strength from the possession of the weapon supreme of mass destruction. For three decades the world has come to realize that the best nuclear policy is never to use the weapons. The holding of mutual hostages, not the active employment of weapons, is still the best policy devised, crazy as it is! Every scheme to make it easier to start nuclear warfare at a "small" scale turns out to contain a terrible but quite natural flaw, that whenever nuclear war is easier to start it becomes harder to stop as well. In this book we say a resolute *no* to every venture of counterforce or preemption, to the whole insecure posture of the easy use of some nuclear weapons. The image of readiness under reason which goes with a secure deterrent, underseas and safe, will protect us better than any clever plans for tradeoffs of explosion for explosion, innocent bystanders meanwhile at their usual peril.

The proponents of more nuclear weaponry like to observe that if their scenario is in error, all that is lost is a few billions (or tens of billions) of dollars. But, they say, if the moderators of nuclear arms are in error, our nation may face unremitting and irresistible pressure for some repugnant accommodation. Given that choice, the usual answer has been obvious. Since the leaders always trust in their own prudence and reason anyhow, there is little risk in nuclear expansion. So they say. But the world cries witness that this short-run claim is in the long run wrong. Step-by-step the weapons improve and grow, while the earth's surface gets no larger at all. Step-by-step we have all slipped under such mistaken "realism" from danger to new danger. Every step upward in potential damage, and in speed of response, has cost us genuine risk, the risk of unreasonable or mistaken leadership, of an error in judgment which for once goes too far. Such is the clear history of warfare. The spring has been wound up, tighter and tighter each decade. It is not the billions wasted which are the greatest real cost, but the growing never-ending escalation of technique and the growing complexities of the tautened balance. The evident risk is that the overwound spring will one day break, and the nations with it.

The debate should not center now on adding more missiles on land, fixed or mobile, nor on improving the delicacy of their triggers. Rather, we should be discussing how fast and how far to dismantle the obsolescent, targetable silos we now have, and by how much and how quickly we might be able to reduce the force loading for the secure undersea deterrent we retain. There is the true direction of the lesser risk, and the way out of the present danger.

THE STEPS AHEAD

The arms race grows out of new developments now as much as it once did from size of forces. We believe American security may often be placed at higher risk—rather than at lower risk—by the increase in speed of nuclear response, by a new type of nuclear weapon, or by a gain in missile flexibility, *even when it is our own.* We therefore propose qualitative restraint, in addition to the quantitative reductions set out in Chapter 7.

1. A comprehensive ban on all nuclear weapons tests by treaty, to be negotiated after the United States has declared a moratorium on its own testing. Such a test ban is today verifiable enough to meet any reasonable objections. The United States should moreover announce its intention *not* to be the first to use nuclear weapons, and perhaps should seek such an agreement internationally. These measures are in the first instance diplomatic; in general in this volume we have not undertaken to discuss the diplomatic implications of military proposals, which are, of course, generally diverse and important. But on this issue we dare to transcend the limits of our study.

2. A halt forthwith to the development or deployment of new options for strategic nuclear offensive war. The B-1 bomber was halted in 1977, but the Trident submarine missile system is in construction and a new mobile ICBM called MX is in development. (Probably the new long-range cruise missiles capable of nuclear loads should also be halted.) These new weapons are not needed for a secure deterrent; worse, their presence is not always clearly distinguishable from a force intended for one or another kind of initiation of nuclear attack, a first strike, or a growing sequence of exchanges. (In this respect, Trident is an exception; it is intrinsically useful as a second-strike system, though capable of first strike as well, especially against "soft" targets.) Earlier the U.S. set aside antiballistic missile defense; that was viewed both as ineffective, and as an incitement to increasing and improving the deterrent forces of the opponent. The 1972 ABM Treaty halted one needless race.

Once in the years after Sputnik, around 1960, the experts told us of a Soviet missile gap. The gap was real enough, but it turned out to be the other way: Soviet ICBM development lagged years behind the U.S. As early as 1966 we were told of new astonishing Russian advances in the antimissile defense. Multiple warheads were first claimed essential to penetrate the defense overseas. That too proved insubstantial; the flurry ended with U.S. MIRVed missiles indeed bearing many warheads, even though ABM was absent. Several years later, the Soviets in turn deployed their many-headed missiles. Once again, they have matched U.S. speeding technical initiatives. Where is the gain? In the end, they will likely be able to follow the lead of U.S. technology anywhere it goes, eventually more or less reaching parity; but what sense is there in this unceasing escalation of risk and threat, we two scorpions in a bottle, each year improving our stings?

The United States is safe against rational attack while it possesses an assured undersea retaliatory deterrent. No force at all can make the United States safe against an *irrational* attack. It is *on both sides* fear, self-deception, and the instability which grows out of an unending search for "superiority," which open the larger risks of unreason and of miscalculation. Those are the worst of all threats to American security. The time is ripe to wind down.

PART III
GENERAL-PURPOSE FORCES

9

PRIMER OF
MODERN WARFARE

GENERAL-PURPOSE FORCES:
LAND, AIR, AND SEA

The Department of Defense divides U.S. combat forces into two great categories, the strategic and the general-purpose. The strategic forces, described in Chapters 6, 7, and 8, include long-range nuclear bombers and missiles, intended to strike at the homeland of major enemies, and short-range interceptor aircraft and missiles, intended to defend U.S. home territory from nuclear bomber attacks. All other U.S. combat forces are covered by the general-purpose category; their small arms, artillery, short-range missiles, and aircraft are armed with high-explosive ammunition. Most of the aircraft and a number of the missile and artillery pieces can, however, also be equipped with nuclear warheads. This gives the general-purpose forces a formidable stock of so-called tactical nuclear weapons, for use on or near the battle front (see Chapter 12).

It is hard to survey the general-purpose forces without seeing a ghostly vision of World War II superimposed on the world of today. The military posture of the United States is fundamentally unchanged, fixed by the same sort of economic interests and military alliances and, above all, by the two oceans. Of all foreign military forces, only the Soviet intercontinental missiles are unhindered by the Atlantic and the Pacific oceans. Apart from a Soviet missile attack, the borders of the United States are not now plausible scenes of land, air, or sea combat on a large scale. Not needed at home, the U.S. general-purpose forces are intended mainly to serve overseas. They are troops intended to engage a land foe in some other country and naval forces intended to open the ocean to American and allied ships and deny its use to others. Contrast this with the case of India

THE STRENGTH OF THE WORLD'S GREAT FORCES

	Armies	Navies	Air Forces
China	3,250,000	300,000	400,000
USSR	1,825,000	500,000	1,500,000
India	950,000	45,000	100,000
USA	790,000	730,000	570,000
Vietnam	600,000	3,000	12,000

FIG. 21

Among the five largest armies in the world, the U.S. Army ranks fourth. But the Army is less than half of total U.S. military forces, a unique circumstance. (Compare Figure 8 for some sense of the reason behind the very large size of the Soviet Air Forces.) If the next largest forces were listed, they would include Pakistan, the two Koreas, France, and then Taiwan, Turkey, and West Germany, the last three very close in numbers. (Figure 5 looks at the whole world military, but in less detail.)

MILITARY BALANCE 1977–78, national entries and Table 3.

or China, Brazil or Japan, Germany or the U.S.S.R.—all of which have conventional forces oriented to the defense of home territory or operations in contiguous areas—and the major postulates of U.S. general-purpose planning become clearer, as does the view of that plan which many must hold in other countries. For the United States still prepares to project its power across the oceans to distant regions where allies and adversaries confront each other. (See Figure 21.)

The potential overseas threats of combat of the general purpose forces shape the discussion in Chapter 10. The regional presentation there derives not only from the importance of geographic factors, but also from the fact that where you fight determines whom you fight, how you fight, and for what goals. When considering the role of U.S. general-purpose forces in Europe, the Middle East, and the Pacific, we take into account the forces and interests of potential opponents who are major military powers—the U.S.S.R. and, secondarily, China. In the rest of the world (Latin America, Africa, and South Asia), which we treat under the rubric of the third world, we show that U.S. military engagements would involve countries with far less military power than the United States.

Before turning to the threaters of combat, we provide in this chapter a brief introduction to the nature of modern nonnuclear warfare on the ground, in the air, and at sea. Describing the most capable U.S. and foreign forces, we begin with land combat. This includes mechanized ground forces which are maintained mainly to deter or fight a war in Europe—the only major theater of land combat the United States can expect to prepare for. It also covers tactical air forces which are maintained to support ground troops and to conduct independent operations in Europe and the Far East (mainly Japan and South Korea). We then look at modern navies, focusing on the U.S.–Soviet competition to control access to the sea lanes and their potential for suppressing coastal navies to undertake blockades, offshore bombardment, or amphibious landings around the globe.

(i) Land Combat: Ground and Air Forces

Human beings live on land, raise their crops and build their homes on land, and organize their societies on the land surface. Military control of territory is in some sense sovereignty: land warfare alone insures that possibility. The basis for this power—the power which grows out of the barrel of a gun—is the organized foot soldier. Nowadays the soldier has much mechanical assistance. He has weapons, light and heavy, from hand-held rifles to long-range rockets and cannon; and he has vehicles of every sort. Trucks and helicopters transport him and supply his needs. Specialized vehicles move him freely on and off the roads, and tanks and other armored vehicles may protect him from widespread forms of attack by small arms. Overhead the foot soldier needs support in the air to prevent enemy aircraft from attacking and escaping unharmed, doing terrible damage to men, supplies, and transport. These factors, familiar since World War II, are not alone in determining the outcome of battle. The infrastructure (roads and buildings), the terrain, the climate, and the tactics can make great differences. Perhaps no factor matters more than the civilians on the ground and their active and passive sympathies, at least for the very long run.

In the U.S. military structure, land-combat forces are held mainly by the Army and the Air Force (Figures 22 and 29). The Air Force provides air support to the ground troops, though for the

THE STRENGTH OF THE US ARMY

Forward Units

17 ⅓ Divisions infantry tanks artillery engineers headquarters close support	
	290,000
Combat Supplement engineers artillery air defense signals electronic warfare additional support	
	180,000
	470,000

Supporting Forces

Base Operations, Maintenance	46,000
Command & Intelligence	63,000
Trainees & Instructors	140,000
Medical Services	30,000
In Transit & Misc.	41,000
	320,000

TOTAL, Active	790,000
Plus the Reserves	590,000

FIG. 22

The seventeen combat divisions of the U.S. Army add up to the 290,000 men shown. To them is added a combat supplement held under higher command, expected to amount to some 11,000 men for each division, or 180,000 more. These are the forward-zone forces now active. In time of full mobilization the supporting forces serve mainly as a framework both to fill out the divisions with additional tactical support increments from the part-time trained Reserves, and to create and service entire new divisions. The wartime manpower for a division would amount to some 45,000, in the three segments described, after a couple of months in the line. This total force is called the Divisional Force Equivalent; informally, the "division slice."

See text notes for the start of Chapter 11.

closest support the Army has its own attack, liaison, and transport helicopters and light aircraft. The Air Force has the responsibility for "airlifting" troops and equipment overseas, and it has medium-sized transport aircraft for use within the theaters of combat.

Among the aircraft which may be available to support land warfare must also be numbered the planes aboard the Navy's aircraft carriers. If free of enemy aircraft and missile attack, these planes can perform missions many hundreds of miles inland. Finally, the Navy also has a small army of its own, the Marine Corps, which has air-support aircraft comprising Army, Air Force, and Navy types. Created in the eighteenth century for ship-to-ship hand combat, the Marines have in this century become specialized forces for beach landings.

Land-combat forces are not as easy to describe concisely as naval or strategic forces. They do not center on a few massive pieces of investment—a few dozen submarines or 200 surface ships or 1,000 large rockets. Rather, they occupy one or two million people with hundreds of thousands of pieces of equipment. Army equipment is more comparable to the 10 million trucks in the U.S. civil economy, or to 100,000 civil aircraft, than to the few ships, strategic bombers, or missiles of the other main types of warfare. U.S. tanks number in the tens of thousands and the tactical aircraft are almost as numerous.

In naval and strategic warfare, the U.S. forces have from time to time dominated all powers put together. But the United States never has and probably never will so dominate the field of land combat. The world is populous and to create an army requires only a modest investment per soldier. This guarantees against a marked U.S. predominance on the ground for the foreseeable future. Indeed, the main aim of U.S. land-combat planning has often seemed to be the substitution of capital investment for manpower, to raise the productivity of a soldier by his equipment, as is the claim of much civil production.

Figure 21 shows the manpower in the organized active forces of the main peacetime military establishments in the world today. (Compare also Figure 5.) It is obvious that large numbers of persons are under arms, and that no small number of these are allied with the United States. Relative to its population, the United States is by no means unmobilized, but there is not the large U.S. lead in numbers that exists for major warships or nuclear weapons.

Since 1973, the U.S. armed forces have operated as part of the competitive labor market. A multimillion-person force is raised without a legal requirement on citizens to serve. Formally, this represents a return to the pre-1948 practice in this country. What is perhaps more important, however, is that it is the first time that a large standing army has been maintained in the United States

FIREPOWER: ABOUT SOME WEAPONS

	Guns, with passive projectiles	Reach	Missile launchers, with active projectiles	Used against:
fired from shoulder	Rifle Machine gun	500 yds		men
			Anti-aircraft Redeye heat-seeking	soft-skinned vehicles
			Anti-tank Dragon wire-guided	
	Mortar		Anti-tank TOW wire-guided	tanks
fired from a vehicle	Gatling gun: Vulcan anti-air, 20 mm	1 mile		
	Tank with 105 mm gun			
		2 miles	Anti-air Chaparral heat-seeking	aircraft nearby
	☢ Medium field artillery 155 mm howitzer, M109			roads, troops in light shelter
held beyond division		10 miles		landing strips, artillery
			Surface-to-air Improved Hawk radar-guided, homing	more distant aircraft
	Heavy field artillery 175 mm gun, M107			fortified positions
		60	Surface-to-surface, ☢ Lance; inertial guidance	railroads
		400	Surface-to-surface, ☢ Pershing; inertial	transportation centers
				depots

☢ nuclear capable   ☢ nuclear

without a draft. The large force fielded in World War II, about six-fold greater than today's, was of course based on conscription. In comparison with the time when U.S. forces last depended on the labor market—in the years between World War I and World War II—the present forces are larger by a factor of 10; allowing for the growth in population, they are still six times larger. There are important questions about this change—political, social, and narrowly military as well. They go beyond the scope of our work. We will assume that the forces raised will be competent and adequate in strength to do their jobs.

GROUND TROOPS

World War I found Europe cut by trenches into two sides which engaged in static mutual slaughter for years, infantry masses going over the top into disaster, against machine guns and artillery, first on one side and then on the other. In World War II, the armored column led by tanks released the combatants to roll over the land in a more fluid way, pitting mobile armor against shell and bullet. The tank remains the center of modern land combat, again taking Europe for the standard.

Modern tanks look not very different from Patton's tanks or the

---

FIG. 23

The more prominent weapons of today's soldier are listed here in rough order of their reach and weight. The scale of distance and the classes of targets are meant only as rough guides.

A mortar is a simple, short-range piece of artillery, usually handled by a few infantrymen themselves. A heavy mortar can lob a twenty-pound shell high over some hill and a mile or two away.

Soon even guns may have active projectiles, capable of some guidance, especially as they near the target. All weapons change as technology and tasks change. For example, *Redeye* will soon become the more reliable and less easily fooled *Stinger*; *Chaparral,* itself derivative from an older air-to-air missile, will be replaced by the *U.S. Roland*; *Improved Hawk* will slowly give way to *Patriot.* All these replacements promise higher performance, but are not changes in broad type.

The M-109 with its ten-mile range is the smallest weapon now in use by the United States which is nuclear-capable. The self-propelled artillery pieces are lighter and much less well armored than the tanks they resemble; they plan to stay well to the rear.

EXAMPLES OF COMBAT VEHICLES

	Weight, Protection & Crew	Typical Weaponry	Weapon Reach & Impact
Jeep with TOW missile launcher	1½ tons soft skin 2-3 men	TOW anti-tank missile, wire-guided	2 miles ✹ armor-piercing warhead
Attack Helicopter with TOW	3 tons bullet-proofed 2 men	8 TOW	2 miles ✹ armor-piercing warhead
Armored Personnel Carrier – M113	10 tons bullet-proofed 2+11 men	11 men with rifles: APC has machine gun, other options	500 yards ✱ ✱ ✱ ✱ ✱ ✱ ✱ ✱ ✱ bullets
Main Battle Tank M 60	60 tons heavy armor 4 men	105 mm: 4 inch gun	1½ miles ✹ high explosive or armor-piercing
Self-Propelled Cannon M109	25 tons some armor 6 men	155 mm: 6 inch howitzer	10-15 miles ✺ high explosive

☢ nuclear capable weapon system

FIG. 24

A few of the vehicles in support of the combat soldier are described in this chart. The figure should be examined with the previous figure in mind. Note the wide and interrelated variety of the weapons.

FIGS. 23 and 24

Assembled from standard references on missiles, armored vehicles, guns. JANE'S WEAPONS SYSTEMS, JANE'S POCKET BOOKS, the infantry manuals of Chapter 11, M109, from tail to gun muzzle, is 31 feet.

German Panzers. They are larger, faster, and fire more frequently and accurately, with a cannon range of several miles. They move at 30 to 40 mph, covering a few hundred miles between refueling stops. The infantry keeps pace in trucks or in light fast armored vehicles of its own, ten men or so to a vehicle. These are the armored personnel carriers.

The basic units of land combat combine under one commander in teams which include the several main kinds of weapon. Such a team of combined arms is called a division. In the American forces a combat division is formed of about 16,000 to 17,000 men (but each is supplemented with many other support units, so that one division "slice" is around 45,000 men in the forward zone). These are organized, as a rule, into groups of up to 1,000 men, more or less, called battalions. Each type of battalion is trained and equipped for a particular weapon set and its use: infantry battalions in armored transport with rifle, machine gun, grenade, and mortar; artillery battalions, with cannon and missile launcher (self-propelled on tracks or towed on wheels), in armed helicopters and in tanks, both heavy for attack and lighter and faster for scouting and patrol (retaining some of the tradition and language of the horse cavalry). Infantry groups have plenty of specialized portable armor-piercing and guided missiles as well, for attacking enemy tanks and aircraft (Figures 23 and 24).

The combat forces are heavily supplemented by service groups. These include signals; engineers; radar and missiles for air defense; fuel, food, and munition supply; medical teams; and of course the special units for command. Over all this there fly the tactical combat aircraft, some after enemy aircraft, some seeking concentrated targets in the rear, others sent directly against tanks, strong points, or groups of men. Still other aircraft and many helicopters, a little out of the range of fire, are devoted to moving men and arms quickly and flexibly, even in areas with dense road networks.

The armored divisions—men, cannon, and tanks—roll forward at high speed, crossing a whole country, it may be, in a few days (Figure 25). Against them is a similar force, plus constant air attack against the tanks and their supplies. Tank forces and low-flying aircraft have so far been seen as the best antitank forces. The defense tries to have prepared strong positions, especially at passes and river crossings. In the short- to medium-term, the tactical outcome seems to rest on who gets first to what points with what forces intact. Yet after an effective projection of concentrated power, the decisive question remains how long a forward force can

THE HOLDINGS OF AN ARMORED DIVISION

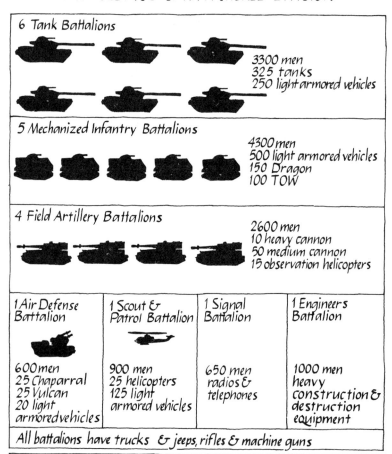

6 Tank Battalions

3300 men
325 tanks
250 light armored vehicles

5 Mechanized Infantry Battalions

4300 men
500 light armored vehicles
150 Dragon
100 TOW

4 Field Artillery Battalions

2600 men
10 heavy cannon
50 medium cannon
15 observation helicopters

1 Air Defense Battalion	1 Scout & Patrol Battalion	1 Signal Battalion	1 Engineers Battalion
600 men 25 Chaparral 25 Vulcan 20 light armored vehicles	900 men 25 helicopters 125 light armored vehicles	650 men radios & telephones	1000 men heavy construction & destruction equipment

All battalions have trucks & jeeps, rifles & machine guns

Plus Specialized & Support Units

Division Headquarters & 3 brigade HQs
   550 men   10 helicopters   10 TOW   10 Dragon
Military Police   200 men   5 Dragon
Intelligence & Electronic Warfare   200 men
Aviation Company   100 men   10 helicopters
Support to mend & feed men & machines   2300 men

TOTAL   16,600 men, an $800 million investment, $400 million per year hold out against the challenges mounted by defenders.

The U.S. divisions equipped with many armored vehicles are popularly called "heavy" divisions. They are particularly aimed at the theater of war in Europe, where armor is also concentrated in the forces of the Warsaw Treaty Organization. In Europe, the land is widely cultivated and accessible, roads are plentiful, and most nations are heavily armed. Of the lighter U.S. divisions, with little heavy armor, some are straight infantry, moving mainly by rail or truck or even on foot; one is specialized to travel by helicopter or fixed-wing airplane; and one is trained to drop by parachute. There are also three light Marine divisions, which are specialized amphibious units. Their training and equipment fit them to land from the sea at places where there is no available port, over the beach, as once was done in Normandy and the Pacific.

All of the various light divisions are more easily transported than the heavy divisions, and they are also considerably cheaper as investments. The armor of one heavy division weighs tens of thousands of tons and costs hundreds of millions of dollars. The light divisions play a secondary role in combat against heavy armor. They claim the skills to fight air-dropped, on beaches, in mountain lands, even in cities. Their strength is aimed at threaters where the opposition is without armor in quantity—that is, typically in the third world.

FIG. 25

The chart reports the make-up of an armored division of the U. S. Army. Such a powerful, varied, and expensive organization is the current basis for ground warfare in the heart of Europe.

Much of the structure is common to all ground warfare. The division integrates all the major types of arms: infantry soldiers, tanks, and artillery. The fundamental unit is the battalion, an organization of some 600 to 1,000 men, a team with complementary skills around some particular kind of weapon, say tanks or artillery. It is the mix of different battalions which differs most from division type to division type.

Armored division structure is now in flux. The U.S. Army is experimenting with a new organization for such a division, in response to the growing importance of precision-guided weapons both for tanks and anti-tanks. The new structure is likely to favor independent operations by smaller groups of tanks over tanks in mass.

Prepared from the infantry manuals cited in Chapter 11, text notes. For experiment, see, for example, *The New York Times,* March 26, 1977, p. 26, an article by Drew Middleton.

TWO DIVISION TYPES COMPARED

	 "Heavy" Armored Division	 "Light" Infantry Division
Men	16,600	16,600
Battle Tanks	324	54
Helicopters	64	160
Artillery Pieces	66	76
Light Tracked or Armored Vehicles	1030	240
Trucks or Jeeps	2800	2900
Anti-air Missiles Guns Shoulder held	24 24 60	24 24 45
Anti-tank Missiles Shoulder held	135 250	160 250
Rifles & Machine Guns	15,000	15,300

FIG. 26

Here side by side are summary descriptions of typical heavy and light U.S. divisions. Intermediate types exist, some with special functions such as paratroops. The differences should be plain upon line-by-line comparison. In rugged terrain or in other hard going the infantry division can make its way, if required, by foot.

The main difference is quite literal. A heavy division with its equipment simply weighs more than a light division: about twice as much. Moreover, its many outsize units—the heavy tanks and guns—weigh together more than five times as much as their few counterparts in the infantry division. The implication for airlift, or even for road or sea haulage, is obvious.

MANPOWER FY 1978, p. V-5. MILITARY BALANCE 1977–78, Table 8, and other infantry sources. Weight calculation by authors, with thanks to Ron Siegel.

Heavy divisions are moving pieces on the chessboard of Europe. They attack, defend, and counterattack. Their overall firepower is remarkable (Figure 27). The new mobility and the amount of accurate explosive power (even without nuclear weapons) are what distinguish present armies from their World War II counterparts. The experience in the Korean War is of little help in anticipating what combat with today's mechanized forces might be like, for Korea was fought with World War II weapons. Even Vietnam is not particularly relevant, for there was no technological parity there. The 1973 Yom Kippur War in the Middle East offers the only paradigm of modern land warfare, though it was fought on a small scale in time and size. Its density and intensity were notable. There is a widespread feeling, perhaps not fully documented, that the whole cycle was enormously speeded up. Losses in men and machines, the profligate expenditure of explosive weaponry, seemed more intense than in the past. The carcasses of thousands of costly fighting machines still litter the Sinai Peninsula—from a three-week battle!

There are portents of real novelty here. The Egyptian Army, for example, had a number of lightly armored trucks—Soviet standard equipment since 1965—each bearing a half-dozen small guided-missile launchers. The missiles—20-pound, wire-guided antitank devices with the NATO code name "Sagger"—travel a mile or two at subsonic speed. Within a space of two hours on one day, 130 Israeli tanks had been knocked out of action, at a cost ratio of some hundreds to one.

The everyday hand calculator shows that electronic control has become remarkably compact, effective, and cheap. Increasingly, these same features characterize modern, short-range missiles on the battlefield. One or two soldiers can carry a Sagger, or a similar American or European homing missile, and fire it with a good chance of bringing down a jet plane or stopping an armored tank. The weapon costs under $5,000 and is sure to become cheaper, lighter, and more effective; a laser beam may replace the wire. Its targets cost as much as a thousand times more and are becoming dearer. Of course tanks can become timid, emit concealing smoke and crouch more warily in the folds of the ground. But as the electronic chips in them grow more subtle, the missiles too will improve: They will be harder to fool and harder to evade. Tank tactics everywhere are already being modified; the U.S. Army is now evolving changes in its tank division structure. Everything suggests that once individual soldiers, operating singly or in pairs, are able to fire a cheap, precision-guided missile at a tank or plane or helicopter, the occasional hit to be expected will swing the balance decidedly to the defense.

In addition to missiles, a number of automatic or semiautomatic aircraft in use or in active development may come to assure functions now carried on by tactical aircraft. Generically called remotely piloted vehicles or RPVs, these devices may patrol the battlefield, make photos or carry other sorts of sensors, and even engage ground targets or aircraft, usually under remote control by operators on the ground or aloft. It is premature to see them, as a whole, as major combat weapons, but they might become highly successful in some special cases.

Some observers predict the coming of fully automated combat to land warfare—the so-called electronic battlefield. Computers connected to a network of devices survey and bug the countryside, dispatch pilotless automatic armed planes, and fire the weapons, to caricature the prophecy. The machines alone suffer, while men program and supervise without injury. This seems far away, not for present concern. In Vietnam, where it was tried in embryo against an enemy with few electronic countermeasures, it was not particularly successful. It will require a tolerant environment (a one-sided engagement) for some time to come. Nevertheless, one aspect of its development, the gradual growth of numerous, cheap hand weapons able to damage expensive, fast machines with small probability but high payoff, does seem likely.

For this reason and others, an attack by an adversary only marginally stronger in number or technique looks less and less militarily attractive, thinking a decade or so ahead. As missiles continue to improve, an attacker must commit what are rather well-known weapons (tanks and aircraft) into new surroundings, with threats that have not been evaluated in battle. As shown in World War II, victory is not always to the well-prepared. The Germans were splendidly prepared, as were the Japanese, but they did not win. Among the risks of a policy of heavy peacetime military preparation, such as that pursued by the United States, is the risk of a bad guess on the nature of future combat. A more defensive posture, which does not rely on the tactical offensive (taking the battle to the enemy), may turn out to have strictly military value in a time of technical change. It is also more stable politically. It does not have as much tendency to spur arms races, to induce the outbreak of "preventive" wars, or to produce avoidable damage in wartime.

To forecast the outcome of a ground campaign is no easy job. The common methods of calculating indices of firepower, weight, speed, and so on are doubtful guides. It is all too easy to strike numerical balances and draw conclusions of no value. Training, motivation, command, even weather must count for a good deal. Then there are nonmilitary factors. How many modern Europeans, for example, east or west of the Elbe, will fight well in an offensive strike far beyond their traditional borders? As long as the force balance remains in some doubt, the idea of deterrence emerges as strong. We do not believe that an army which may be leading in one or another index but which does not enjoy manifest superiority across the board is likely to try an aggressive run against a powerful force more or less its equal. The day of the defense, if not yet here, may well be dawning.

TACTICAL AIR FORCES

Troops, guns, and tanks tend to fight against their mirror images. The same tanks can be used offensively, to intrude into enemy lands, or defensively, to guard the capital. The demands of air combat, in contrast, tend to force distinct designs on aircraft intended for differing tactical roles. Tactical airpower can be divided into two groups: planes that attack ground targets and those that fight other airplanes. Each group can be further divided into a long- and short-range component. The following sections describe these functions and the aircraft suitable for performing them, concluding with a brief note on special-purpose aircraft. (See Figures 28 and 29.)

The antiground aircraft.

The mission in which planes fly across the front line of battle to attack surface targets at long range is called interdiction. Its purpose is to reduce the ability of the enemy to bring force to bear at the front; this is distinguished in U.S. doctrine from strategic air attack, which involves strikes against people and production facilities deeper into the homeland. Targets for interdiction include bases, airfields, road junctions, pipelines, fuel depots, ships, ports . . . the list is long. The intent may range from aggression in support of an invasion, at one extreme, to counterattack in support of a

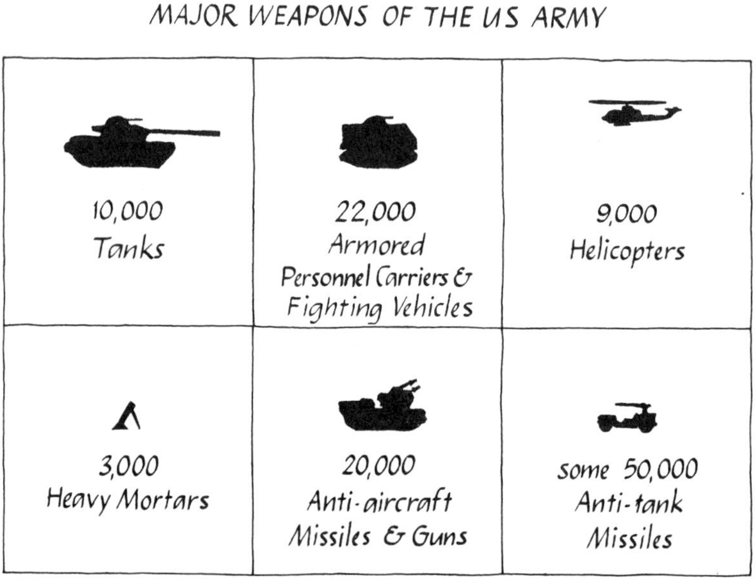

MAJOR WEAPONS OF THE US ARMY

10,000 Tanks	22,000 Armored Personnel Carriers & Fighting Vehicles	9,000 Helicopters
3,000 Heavy Mortars	20,000 Anti-aircraft Missiles & Guns	some 50,000 Anti-tank Missiles

FIG. 27

The Navy counts its ships of any one type in the score, the Air Force numbers aircraft of a class in hundreds, but the U.S. Army holds its major weapons by the ten thousand. It is exactly this multiplicity of armed and organized men on the ground which is the essence of the Army. The Army helicopters include many heavily armed anti-tank and anti-personnel craft, but most of them are for observation, liaison, or transport. The missiles are mostly the shoulder-held sorts, both anti-air and anti-tank.

These inventories are being improved and expanded every year.

JOINT CHIEFS FY 1978, p. 63, and entries for missiles in PROGRAM AC-QUISITION FY 1976.

defensive position, at the other. However, the range, speed, and armament of aircraft equipped for this role make preparation for it ambiguous. Potential enemies are apt to regard a growing inter-dictory force as an aggressive threat.

The attack aircraft suitable for interdiction tend to be very costly. They must go far with a heavy load of munitions, so they are large and heavy. At the same time, they are on the tactical offensive, often flying over enemy land, close to well-prepared and massed defenses.

For this they must be fast and have means of self-defense. This combination of requirements places high demands on performance: demands which in the event have often been unrealizable. Moreover, the interdiction mission is becoming increasingly difficult, risky, and costly as surface-to-air missiles improve. Vietnam showed another sort of obstruction to the mission: a failure less from technical deficiency than from the simple and resilient nature of the enemy organization on the ground. On the other hand, the strike by Israel against Egyptian aircraft on the ground in the 1967 War exemplifies a successful interdictory attack. Like the Japanese attack on Pearl Harbor, however, this illustrates more the one-time advantage of a preemptive attack at the outset of hostilities than the requirements and conditions of an extended interdiction campaign waged in the midst of a war.

The U.S. Air Force and Navy have long emphasized interdiction in their tactical aircraft inventories. The most recent aircraft designed for the role are the Air Force's costly F–111 and the Navy's F–14 Tomcat. The heaviest supersonic combat aircraft ever built, both of these planes have experienced technical shortcomings never fully resolved. The smaller but versatile and long-lived F–4 Phantom, in contrast, has for decades been phenomenally successful in the same role. The F–4 probably brought this type of aircraft close to the optimal combination of features, although its range and weapons load are less than that of the F–111. Over the years since it was introduced in 1958, about 4,000 F–4s have been produced for the U.S. Air Force and Navy, as well as about 1,000 for 10 other countries; it is still in production.

In contrast to the long distances typical of interdiction, close air support aircraft attack enemy ground forces near their own ground troops and in support of them. The aircraft can be relatively slow, of light-to-medium weight, able to bomb and strafe infantry, cannon, tanks, and strong points. Armed helicopters, as well as fixed wing planes armed with antitank weapons, can perform this mission, not looking into the muzzle of a gun or peeking over a wrinkle in the ground but seeing down over a whole panoramic section of the battlefield laid out below them. This was the most common form of close air support in Vietnam. Such air cover generally requires at least local or transient air superiority: attack planes, to say nothing of the slower helicopters, cannot survive long if faster fighters are around them in numbers. For this reason, air support is most easily provided in an environment where there is an outmatched foe. That would be true only locally in Europe and it is getting harder

THE CHARACTER OF SOME COMBAT AIRCRAFT

F-111A Anti-ground, interdiction fighter-bomber. Max speed 1650 mph Max take-off weight 45 tons Weapons load 6 tons Since 1967 $11 million 1967 $	Can operate at high altitude – 12 miles – or follow terrain at 100 feet. 562 built. Carries bombs, or cannon and air-to-air missiles. 2 seat, swing-wing, twin jet.

A·10A Anti-ground close support attack aircraft Max speed 520 mph Max take-off weight 23 tons Weapons load 6 tons Since 1977 $5 million	Can "loiter" at low speed for two hours loaded. Carries 30mm Gatling gun (1350 rounds), plus varied load. All sorts of bombs & missiles, to one ton size. 1 seat twin jet.

F-15A Anti-air, air-superiority– "Top of the line" Max speed 1650 mph Max take-off weight 28 tons Weapons load 7 tons Since 1975 $17 million	Highest maneuverability & acceleration. Flies to 13 miles altitude. Carries 20mm cannon, 500-600 rounds; air-to-air missiles. 1 seat, 2 jet.

DC-9 Intercity Passenger Jet for comparison Max speed 550 mph Max take-off weight 49 tons Payload 13 tons Since 1967 $5 million	Carries a hundred passengers & their baggage. Typical range, 1700 miles. Over 800 built by 1976. 2 crew, 2 jet.

F-4E Multi-use fighter for anti-air, anti-ground plus reconnaissance Max speed 1550 mph Max take-off weight 27 tons Weapons load 6 tons Since 1964 $5-6 million	Most successful; a dozen models. 4740 built by 1976 for USN, USAF, plus 10 other countries. Carries 20mm cannon, air-to-air missiles, many other options. 2 seat, 2 jet.

F·16A Anti-air fighter Max speed 1300 mph Max take-off weight 16 tons Weapons 5 tons Since 1978 $8 million	Highly maneuverable. Carries 20 mm cannon, 500 rounds, air-to-air missiles. 1 seat, 1 jet.

everywhere, as surface-to-air missiles improve and become less costly. Indeed, such missiles even gave the U.S. Air Force cause to be wary flying over the jungles of Vietnam and the Israeli Air Force concern over the Suez. Despite the missile risk, the Air Force is in the process of responding to a long-time complaint of the Army that its complement of attack aircraft, composed of F–111s, F–4s, and the Air Force version of the Navy's A–7, a light bomber, gave far too much attention to long-range interdiction and far too little to short-range close air support. Thus, the A–7s are being replaced by the new A–10, and Air Force plane especially designed for close air support. There will be a larger proportion of close air support squadrons, at the expense of some squadrons now aimed at interdiction and air superiority missions.

Antiair aircraft.

Fighter aircraft performing air superiority missions attempt to clear the air of enemy aircraft, both enemy fighters out to do the same job and enemy bombers out to attack ground forces. Their weapons are not bombs, but antiair weaponry such as quick and clever missiles, good at target-seeking. Often, these aircraft operate over their own troops and territory, blocking enemy strikes or clearing the way for their own interdiction aircraft. The stresses of "dog-

---

FIG. 28

Five modern combat aircraft are here compared with respect to their important capabilities. Notice the DC-9 present as a familiar comparison; in fact, this aircraft, called a C-9, serves the Air Force as a medical evacuation plane. The costs are steadily rising, comparing planes by weight or any other measure. Some of that is of course the mark of inflation—the years of purchase differ—but most of it is the increasing complexity of the planes, their electronics, and their missions.

Many differences in the aircraft are obvious. The nimble new F-16 turns neatly inside the fast turn of the reliable old F-4; the original photo was published by the proud manufacturer. The F-111 fighter bomber more than twice outdistances the new A-10 combat aircraft, with a range out and back of about 1400 miles with useful load.

Usual aircraft references, as in text notes, plus recent advertisement of General Dynamics, LaClede Center, St. Louis, for the F-16 (*Aviation Week,* Feb. 6, 1978). Also *International Defense Review,* Combat Aircraft (1977).

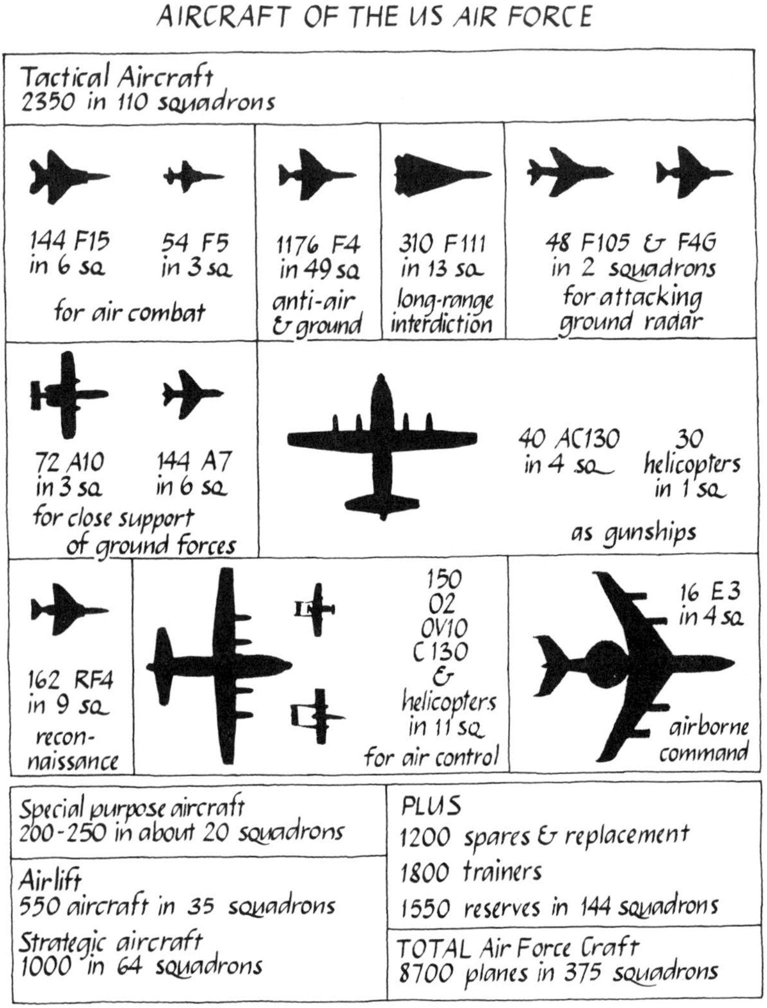

AIRCRAFT OF THE US AIR FORCE

Tactical Aircraft
2350 in 110 squadrons

144 F15 in 6 sq	54 F5 in 3 sq	1176 F4 in 49 sq	310 F111 in 13 sq	48 F105 & F4G in 2 squadrons
for air combat		anti-air & ground	long-range interdiction	for attacking ground radar
72 A10 in 3 sq	144 A7 in 6 sq		40 AC130 in 4 sq	30 helicopters in 1 sq
for close support of ground forces			as gunships	
162 RF4 in 9 sq recon- naissance		150 O2 OV10 C130 & helicopters in 11 sq for air control		16 E3 in 4 sq airborne command

Special purpose aircraft 200-250 in about 20 squadrons	PLUS 1200 spares & replacement
Airlift 550 aircraft in 35 squadrons	1800 trainers 1550 reserves in 144 squadrons
Strategic aircraft 1000 in 64 squadrons	TOTAL Air Force Craft 8700 planes in 375 squadrons

FIG. 29

The active aircraft of the U.S. Air Force, drawn to scale, are arranged here by type, with total numbers and squadron organization. Like an Army battalion or the crew of a middle-sized cruiser, an Air Force squadron is a close-knit unit of some 600 to 800 men. The bigger and more complex the plane, the fewer planes in general can be served by one squadron. The number of planes given is the so-called Unit Equipment, or UE, the number for which all needed maintenance fights" make heavy demands on the design of this type of aircraft. They need high speed and maneuverability, as well as accuracy of antiair armament and strong countermeasures against enemy weaponry. Training and skill with tactics can, however, influence the outcome of encounters as much as these technical factors.

The Navy's F-14 is designed to perform this mission as well as interdiction, with the result that great demands were placed on the design of the craft. An attempt to develop a more powerful engine, to be refitted (at a cost of 3 billions) into existing units to improve the fighter performance of this heavy plane is under study (1978). Another plane of this type is the Air Force's new F-15 fighter, a large, fast plane heavily armed with antiair weaponry. The F-14 and F-15 are both high performance fighters which combine speed, maneuverability, low vulnerability, all-weather use, and con-

---

and flight personnel, weapons, and so on are provided to the squadron. The Air Force always owns more aircraft than the UE number; the remainder allow for aircraft out for repairs, replacements, for loss, training, and so on. The allowance varies with type, in the neighborhood of 30 percent above UE.

The Reserve aircraft are listed as a whole, along with the planes for training and spares. (Airlift and strategic planes are described in detail in Figures 55 and 15, respectively.)

These numbers change. The chart applies to early 1978; we round off many numbers not to overstate the precision. The number of the newest fighter squadrons with the F-15 is fast increasing, while F-4 squadrons are going down, as the Air Force switches to the newer planes. The total number of such fighter squadrons remains near 58, but the types change. The three F-5 squadrons are in fact used mainly as air sparring partners for the bigger fighters; this light aircraft is a reasonable stand-in for the numerous light MiGs of the U.S.S.R. forces. It seemed fair to count them in the mix.

MILITARY BALANCE 1977–78, and the Almanac Issue of *Air Force* magazine, May 1977, throughout, especially for numbers in squadron and for detailed inventory. Unit Equipment is discussed in MANPOWER FY 1978, Chapter XIII. *Air Force* gives an account of the F-5 squadrons. The F-15 changes are indicated in DEFENSE SECRETARY FY 1978, p. 207 and JOINT CHIEFS FY 1978, p. 83.

A view of an Air Force squadron is given in MANPOWER FY 1978, p. XIII-20. A squadron of standard multipurpose F-4E aircraft holds 24 planes, a flight crew of 2 in each plane aloft. The squadron has 60 flying officers, 430 maintenance personnel, 140 handling munitions, 45 for staff and management, 40 for security, for a total of about 720 personnel. The C-130 wingspan is 133 feet.

siderable weapons load and range—and which do so at high cost. The unit price of these superplanes is so high—now $16 million for the F-15 and $21 million for the F-14—that the Air Force and the Navy have been developing smaller, cheaper fighters to mix with them. These planes, like the Air Force's F-16 (expected to cost $6 to $7 million) and the Navy's F-18 (prototypes only) do not fly so far, so fast, or with so much protection and weaponry. Suited to operate closer to friendly forces, these lighter fighters are oriented more to defensive air combat nearer the battle area and less to intrusion deep over enemy territory. They represent a moderation in demand and resemble their Soviet counterparts more than do the F-14 and F-15, which have no peers.

Air defense or long-range interception can be seen as a special case of the air superiority mission. It differs in emphasis in two respects. First, it is aimed at stopping not so much enemy fighters as attack or bomber aircraft, which tend to be less maneuverable and slower (commonly flying at or under Mach 1, the speed of sound). Second, this mission is more closely tied to the tactical defensive. This may mean short-range operations (100 to 200 miles) around the perimeter of a country or it may involve longer flights (500 to 1000 miles) to intercept bombers far from the border or to shoot down antishipping attack aircraft before the enemy aircraft come within the effective range of a ship's air-to-surface missiles.

Aircraft especially designed for interception have not been particularly important in the U.S. force structure, which relies on the long-range carrier-based F-14 air superiority fighter for interception of strikes aimed at surface ships; the dual-purpose fighter-attack F-4 for both Navy and Air Force interception of attack aircraft; and the F-106—a specialized Air Force interceptor aircraft designed in 1958 and now retained in small numbers—for air defense of the continental United States against bomber attack.

In Europe and the Soviet Union, air defense and interception play a much more important role in the tactical air forces. In that area it is more difficult to draw a line between the strategic air defense mission, aimed at intercepting enemy long-range bomber aircraft on their way to attack one's home territory, on the one hand, and on the other hand, the tactical air superiority mission, aimed at intercepting shorter-range attack aircraft on their way to provide interdiction or close air support to their ground forces in a local engagement. The two types of attack overlap, as do the two types of interception. For this reason, the European countries and particularly the U.S.S.R. and Eastern Europe have large numbers of specialized tactical combat aircraft whose primary mission is strategic air defense, but which can also be used in the tactical air superiority role.

Special-purpose and reconnaissance aircraft.

More and more special roles are appearing for tactical aircraft. These include airborne warning by radar and radio; reconnaissance by photographic image and electronic means; electronic warfare and the countermeasures to it (jamming, disruption, or attack of radar installations and the like); and air control aloft, coordinating the operations of warning, reconnaissance, and fighter and attack aircraft. Perhaps as much as one-fifth of the advanced air forces is made up of this type of specialized aircraft, for example, the E–2, EA–6, EC–135, and WC–130. (Compare Figure 29.)

In sum, it is clear that there are two main types of tactical combat aircraft which can perform the various missions. The one is an expensive, high-technology, intrusive sort of plane, which projects power forward at speed. Such a plane may be intended for defense alone, but it can easily be used for aggressive purposes as well, and for that reason it tends to be destabilizing. The other sort of plane is smaller, lighter, shorter-range, slower, and cheaper. This type of aircraft can be procured in larger numbers or save money or do some of both. It offers less risk of technical disappointment or costly vulnerability to technical advances in antiaircraft weaponry, and it gives a clear signal of defensive intention.

The U.S. tactical air forces have emphasized what the Defense Department calls "high-mix" aircraft—the more powerful types, the older F–4 and F–111 and the newer F–14 and F–15 (Figure 29). But for reasons of cost and strategy, the proportions are shifting somewhat to give more emphasis to the "low-mix" A–10, F–16 (and possibly F–18). This is a tendency which we welcome as safer, cheaper, and more stable. There is something of an opposite trend in the Soviet Union, where large numbers of simple, light, short-range planes are gradually being replaced by more capable aircraft. The most advanced of these, not mainly a tactical plane, goes back to about 1970. It is the MiG–25 Foxbat (the number is Soviet, the code name Western). The MiG–25 was designed to be an air defense interceptor, aimed at beating back the U.S. B–70 long-range bomber—a supersonic high-altitude plane which was debated during the 1960s and, ultimately, never built. The MiG–23 Flogger

SURFACE SHIP TYPES

	Weight: displacement Speed Crew	Main Armament
Aircraft Carrier, Nimitz 	95,000 tons 30+ knots 6300 men	100 aircraft
Largest Passenger Liner Queen Elizabeth II  for comparison	66,000 tons 29 knots 1000 crew	1800 passengers & many amenities
Amphibious Warfare Ship Tarawa houses & transports troops & equipment to land over beach	40,000 tons 20 knots 1800 men	30 helicopters 3·5" guns 6 anti-air guns
Destroyer, Spruance  escort & anti-submarine medium size gun & missile ship	8,000 tons 30+ knots 250 men	2-5" guns 1 anti-sub missile launcher 2 helicopters
Patrol Gunboat, Asheville	240 tons 16/40 knots 24 men	1·3" gun 1·40mm anti-air gun 9 machine guns

FIG. 30

The figure presents true-to-scale silhouettes and data on a few important naval ship types. The famous Queen Elizabeth II offers a familiar comparison, small is the Soviet's best air superiority tactical fighter—a plane ten years old, rather like the French Mirage of the same era. Even the recent models of the very successful MiG–21, first introduced in 1960, are claimed to be quite capable. They are not sufficiently advanced, however, to have changed the long-time rule of thumb for expected Soviet-to-Western tactical aircraft losses, which stands at two to one, especially in a defensive context. (See Figure 39 in Chapter 10.)

(ii) War at Sea

More than the land-combat forces, U.S. naval forces are capable of performing many different functions. Their potential roles run the gamut from showing the flag in a friendly port to policing international straits, waging a traditional war at sea, or even undertaking nuclear-bombardment of land-based targets.

Despite this flexibility, naval forces must be judged with reference to their primary missions. The reason is that these primary missions alone exploit the specialized capabilities of the ships; and a pressing and specific need for the special capabilities must be shown to justify their exorbitant cost. At $200 to $1,500 million apiece (compared with $1 million each for tanks and $15 million each for tactical aircraft), ships are by far the single most expensive item of military equipment. Ships are more precious than other types of equipment not only in procurement costs, but also in production time (years rather than weeks or months), crew sizes (hundreds or thousands rather than tens), and crew training costs. Ships are also very limited in numbers—which is not surprising, given these factors. Only about 80 attack submarines and about 200 oceangoing surface warships have been built in the United States since World War II. In addition to their high cost, long construction time, and small numbers, ships are vulnerable to crippling damage by vastly cheaper and more numerous missiles, torpedoes, and bombs. All of these factors increase the stringency with which one must apply the universal military requirement that wartime deployments of naval forces must be "cost-efficient." The value of a ship's

---

next to a U.S. aircraft carrier. The little *Asheville* docs not count as an ocean-going, or bluewater, ship; it is a modern coastal patrol boat, the type found widely throughout the world in smaller navies. The U.S. Navy has not much need for such craft.

target or mission must be commensurate with the ship's own value
and with the risk it runs of being successfully attacked.

PRIMARY MISSIONS OF THE U.S. NAVY

There are three primary missions of the Navy's general-purpose
ships (that is, all warships except the strategic missile-bearing
submarines described in Chapter 7):

1. to provide floating air bases for light and medium bomber
   aircraft, which can attack enemy ships and land-based
   targets up to several hundred miles inland in areas distant
   from allied or uncontested air fields;
2. to transport troops and their arms (principally the Marine
   Corps) for amphibious assault landings on overseas beach
   areas held by hostile forces; and
3. to provide antiaircraft and antisubmarine defense of war-
   supply shipping from the United States to Europe in the
   event of a prolonged nonnuclear war with the Soviet Union
   in that theater.

The air bombardment mission is the task of *13 aircraft carriers,*
with a combined capacity to deploy over 1,000 combat aircraft.
Landings on hostile beach areas are to be launched from a fleet of
about *63 oceangoing amphibious landing ships,* together capable
of deploying one and one-third Marine divisions together with their
tanks, helicopters and other combat equipment, air and ground
support forces, and beach landing craft. (A total of about 40,000
troops can be transported in U.S. Navy amphibious ships.)

The third function, protection of wartime shipping to Europe, is
performed by two types of force. The first is composed of escorts or
short-range antiair warfare (AAW) and antisubmarine warfare
(ASW) defense of convoys. These "point-defense" escorts are
destroyers and frigates, whose weaponry is confined almost
exclusively to AAW and ASW types. Their equipment and speed
are suited for short-range (10- to 20-mile) operations around the
perimeter of a group of unarmed transport ships during transoceanic
crossings. This is a last-ditch defense, supplemented and preceded
by wide-area sweep antisubmarine forces which patrol the North
Atlantic for Soviet antiship armed submarines. The wide-area
search ASW forces are composed of *70-odd nuclear-powered
submarines* (often referred to as hunter-killer submarines, designed
to seek out and destroy other submarines, as distinct from antiship

SUBMARINE TYPES

	Size: displacement submerged	Speed submerged	Crew size	Armament
US Lafayette Class  — Ballistic Missile Submarine	8250 tons	30 knots	145 men	16 ballistic ☢ missile tubes 4 torpedo tubes
US Los Angeles Class  — Anti-sub Submarine	6900 tons	35+ knots	127 men	4 torpedo tubes which launch anti-sub missile SUBROC☢ anti-ship cruise missile Harpoon
USSR Echo Class  — Anti-ship Submarine	5600 tons	22 knots	100 men	8 cruise missile ☢ surface launchers 10 torpedo tubes

☢ nuclear weapon
☢ nuclear capable weapon

all of these subs are nuclear powered

FIG. 31

The new element of today's great navies is the big nuclear-propelled submarine. Such a sub, whether it carries strategic missiles or only anti-sub or anti-ship weapons, is free from need to surface for weeks on end. The crew's tolerance for confinement usually sets the time limit. The Soviet Echo-class boat, for example, is much less roomy and comfortable than are the larger subs of the U.S. Navy. The U.S. anti-sub submarines are newly gaining a secondary anti-ship function with the new anti-ship missile, *Harpoon,* which can be launched through the torpedo tubes while the boat remains underwater. The Echo cruise missiles require surfacing for launch.

FIGS. 30 and 31

Data from JANE'S SHIPS 1976–77. *Queen Elizabeth II* from Cunard Line. The *QE II* measures 940 feet long. SSBNs of *Lafayette* class are 425 feet long.

A CARRIER TASK FORCE &
AN UNDERWAY REPLENISHMENT GROUP

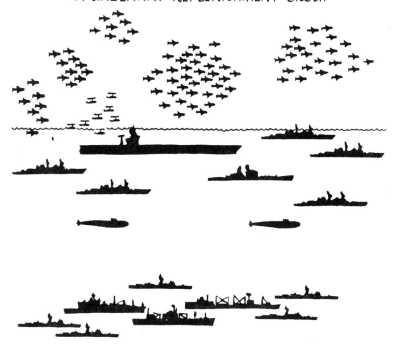

FIG. 32

The seascape offers an impression of a carrier task group at sea with its underway replenishment group. (The 100 or so carrier planes are aloft mainly for show.) Around the carrier for many miles steam its protection from surface missile and submarine attack: one cruiser, five destroyers, and two anti-sub submarines. Such a group would be typical of a voyage far from land attack. A carrier needs much: food for the men, fuel for the planes (even if the carrier itself is nuclear-propelled), ammunition and missiles. Those supplies must be brought to it (and to its screening ships) every couple of weeks in normal action, or the big ship itself must travel to its distant home port. The underway replenishment group is expensive and vulnerable—its ships are fast for cargo vessels so that they can transfer their supplies while donor and recipient are underway at speed. This replenishment group has three ships, one oiler, one ammo ship, and one for combat stores. Its own protection screen includes five frigates.

Here one sees much of the role of the surface combatants of the Navy; such a carrier task force is both expensive and vulnerable to missile attack today.

"attack" submarines); and *200 land-based antisubmarine patrol aircraft*. These specialized aircraft are equipped with antisubmarine weaponry and sensors—including radio sonobuoys dropped at sea—and are designed to patrol over the ocean for periods up to 16 hours at low speed (under 200 mph) (Figures 30, 31, and 33).

In addition to the forces mentioned so far, there are about *165 cruisers, destroyers, and frigates* whose main function is to provide antiair and antisubmarine defense for the aircraft carriers and the amphibious landing ships. Some do this directly, by escorting the carriers and landing ships in small "task forces." Others escort the unarmed support ships which are used to resupply the carriers and other surface ship forces. These support ships, which carry ship fuel, aircraft fuel (for the carrier aircraft), ammunition, food, and other supplies, are organized into "underway replenishment groups" which permit the warships to remain on station for months at a time rather than having to return to port at frequent intervals (Figure 32).

The block retirement of large numbers of World War II ships in 1970-76 has left the number of surface escorts (cruisers, destroyers, and frigates) 10 to 20 short of the level normally planned for escorting carriers, amphibious ships, and underway replenishment vessels; and there are in fact no escort ships available for transport convoys. About 30 World War II destroyers now used for Naval Reserve Training could be employed for this purpose, but they will be retired during the next few years. To compensate, the Defense Department has planned a major frigate building program which, over the next decade, will increase the number of convoy escorts by 50 to 60. The current ship-building program will also raise the number of hunter-killer submarines by about half, to somewhere between 90 and 105. Other ship types are being built or modernized at a rate which, allowing for a regular retirement of aged ships (25 to 30 years depending on type and condition), will keep the size of the Navy about constant.

NAVAL FORCES OF OTHER COUNTRIES

In contrast to the case for ground and tactical air forces, the navies of other countries are *not* roughly similar to that of the United States

---

Drawn from account in MANPOWER FY 1978, p. V-13, and ship silhouettes and descriptions, e.g. in *Combat Fleets of the World*, 1976/77, edited by J.L. Couhat, Naval Institute Press: Annapolis.

US NAVAL SHIPS~

	Active ships	Displacement range
US Navy:		
Carriers		
Aircraft Carriers	13	64-90
Major gun & missile ships: 162		
Cruisers	26	8-18
Destroyers	72	3-8
Frigates	64	2-4
Submarines		
Strategic missile subs	41	7-8
Anti-sub, attack subs	68	3-8
Patrol subs	10	2-3
Amphibious warfare ships	63	8-39
Support ships		
Underway replenishment ships	40	17-54
Major & minor fleet support	45+	varied
US Coast Guard:		
Cutters, frigate equivalents	28	1-3
Older cutters & patrol boats	91	small

Displacement is given in thousands of tons

FIG. 33

The ocean-going ships of the active Navy are listed here. There is wide variation in size among ships of a single type. We include the U.S. Coast Guard, because it holds ships of frigate-equivalent size and function, with many more coastwise patrol craft.

Naval aircraft serving 13 carriers
1350 aircraft in 126 squadrons:

144 F14 in 12 sq	168 F4 in 14 sq	144 A6 in 12 sq	324 A7 in 27 sq	about 100 E2 & EA6 in 12 sq
Anti-air	Air & ground attack	Surface attack		Radar & countermeasures

120 RA5 & RF8 in 10 sq	110 S3 in 11 sq	88 helicopters in 11 sq	+ about 150 aircraft in 17 sq
Reconnaissance	Anti-submarine		

Deployed otherwise than on carriers
700 aircraft:

	216 P3 in 24 sq	On other ships & on land
	Land-based anti-sub patrol	400-600 helicopters

Marine Air Wings
900 aircraft in 54 squadrons

144 F4 in 12 sq	60 A6 in 5 sq	60 A4 in 5 sq	80 AV8 in 3 sq	21 RF4 in 2 sq	18 EA6 in 2 sq	54 OV10 in 3 sq
Anti-air & ground	Ground attack		Vertical takeoff	Recon	Counter-measures	Close observ'n

36 KC130 Tankers

425 helicopters in 22 squadrons

PLUS Navy & Marine

800 spares & replacement
1100 trainers
800 reserves

5700 TOTAL Aircraft

in size or composition. Other navies differ substantially in numbers of ships or different types. They also vary in the size of the ships, which determines capacity for fuel, ammunition and other stores, range, maximum duration of tours of duty, and rate of fire. Differences in these factors will determine whether ships are suitable for coastal defense, for operations in enclosed seas, or for duty on the open ocean at considerable distances from home port. In addition, within each sphere of operations, ships' roles may vary with their sensors, armament, and auxiliary support.

Only the United States and the Soviet Union can deploy more than a very small number of ships (two or three) on the high seas in all of the world's oceans. The navies of most other countries are limited to a few oceangoing surface vessels (20 or fewer), and these ships are intended mainly to strengthen the aircraft, patrol boats, and other units comprising their coastal defenses, not to perform combat operations far out in the ocean or overseas. The half-dozen exceptions, notably Britain, France, and Japan, have a somewhat larger number of oceangoing ships and submarines—in the neighborhood of 40 to 50 each. These are to be deployed in ocean areas relatively near their own coasts, acting in concert with U.S. forces in operations oriented against the Soviet Atlantic or Pacific fleets. The next ranking navies include those of some other Western countries—Sweden on the Baltic and Italy and Spain on the Mediterranean—China, and India. China forms the third side of a triangle around the Sea of China, with the U.S.S.R. and Japan on

The Navy has its own tactical air force, mostly carrier-based. It has a special anti-submarine warfare air fleet, some carrier-based, but also with hundreds of big long-range armed anti-sub (and anti-ship) patrol planes based on land. It has in the U.S. Marine Corps an army of its own, with the necessary close support and combat aircraft, plus a large fleet of helicopters, most for transport of Marines, but many armed. Navy and Marine squadrons are much smaller than the Air Force standard, for it is important to have a mix of many types on each single carrier or amphibious force.

The E-2 planes are the carrier's modern "crow's nest," providing early radar warning aloft. They also control the carrier's fighters.

JANE'S SHIPS 1977–78, and Couhat, as above. Also DEFENSE SECRETARY FY 1978, p. 217ff. We believe the active units are well-estimated here, but total Navy inventory of aircraft appears somewhat larger than we estimate, perhaps because the Navy still holds many older aircraft from the larger carrier force some years back. (Inventory material is found in a leaflet of the Office of the Navy Comptroller, *Budget and Forces Summary,* NAVSO P-3523, Nov. 1977.)

the other two sides. Of the Soviet Union's East European allies (Bulgaria, East Germany, Poland, and Romania have access to the sea), Poland has two ex-Soviet oceangoing warships and the others have only large coastal defense patrol craft.

The Soviet fleet is about the same size as that of the United States in numbers of oceangoing combat ships, but the similarity between the two ends there, since the Soviet Navy is structured to counter rather than to parallel the U.S. Navy (Figure 34). The Soviet Union has no carriers to serve as floating airfields for high-performance bomber aircraft; no large, ocean-crossing amphibious landing ships; no overseas allies which it might support with ground troops, to be resupplied by sea-route convoys; and very few underway replenishment support ships, requiring escorts on their way to forward-deployed warships.

The Soviet non-strategic, oceangoing fleet comprises 60 submarines armed with antiship cruise missiles; 100 nonmissile submarines equipped with antiship torpedoes; and about 75 oceangoing surface combatants (cruisers and destroyers), most of which are also armed with antiship cruise missiles. These antiship forces, which have no real counterpart yet in the U.S. Navy, are intended to interdict U.S. aircraft carriers and convoys in the Atlantic and the Pacific—particularly, in the event of a war in Europe, in the Atlantic. The submarines form the first line of interdiction. The surface ships, supplemented by some 40 ships armed with guns, around 100 small frigates and two large cruisers equipped with antisubmarine helicopters to defend surface task forces from submarine attack, form a second line of interdiction. This is closer to Soviet ports in the North Atlantic (above Norway), the Baltic Sea (near Finland), the Black Sea (past Turkey and the Bosporus), and the North Pacific (past Japan), to which they must return for fuel and ammunition resupply. The second line is supplemented by Soviet naval mission aviation, which can put up about 500 medium-range, land-based, bomber aircraft capable of attacking ships with free-fall bombs and air-to-surface missiles. As a final line of defense, the Soviet Union has numerous large and small patrol boats, some with cruise missiles, as well as some mine warfare ships and short-range submarines, to be used in coastal waters.

In addition to its interdiction forces, the Soviet Union has three small aircraft carriers (two still under construction) which cannot launch supersonic combat aircraft, but which carry helicopters and subsonic vertical takeoff-and-landing (VTOL) aircraft that might

US & SOVIET BLUEWATER FLEETS COMPARED

	Number of ships		Total weight millions of tons	
	US	USSR	US	USSR
SURFACE FORCES				
Aircraft carriers	13		1.1	
Anti-sub missile & helicopter carrier Helicopter carriers		1 2		0.1
Cruisers, missile armed Cruisers, gun armed	26	21 10	0.3	0.4
Destroyers, missile Destroyers, gun	38 34	53 34	0.4	0.4
Frigates in Navy Frigates in Coast Guard	64 28	107	0.2	0.1
TOTALS, Surface	203	228	2.0	1.0
SUBMARINE FORCES				
Strategic missile subs, nuclear powered Missile subs, not strategic, diesel powered	41	62 20	0.3	0.6
Anti-sub subs, nuclear	68		0.3	
Anti-ship subs, nuclear		76		0.2
Patrol subs, diesel	10	150+	~	0.2
TOTALS, Submarine	119	320	0.6	1.0
SUPPORT FORCES				
Fast underway replenishment ships	40	20	0.9	0.2
Amphibious warfare ships	63	20	0.7	0.2
Fleet support, major & minor	60	30+	0.4	0.2
TOTALS, Support	163	70	2.0	0.6
GRAND TOTALS	485	620	4.6	2.6

be used for antisubmarines, antiship or close air support duties in an area free of enemy fighters. The U.S.S.R. also has a score of small oceangoing amphibious landing ships and 75 landing craft, which provide the capability for small-scale amphibious landings within a few hundred miles of Soviet borders.

THE CHANGING TECHNICAL BALANCE AMONG SHIPS, SUBMARINES, AIRCRAFT, AND MISSILES

During World War II and the first 10 to 20 years thereafter, the aircraft carrier dominated the surface navy. This meant that after the defeat of Japan the United States dominated the seas, since no other country maintained substantial numbers of carriers. U.S. carrier-based attack (light and medium bomber) aircraft could pick off enemy ships with relative ease, while the carriers could remain at distances many times the ranges of the ships' guns. As a result, there were no challenges to the U.S. Navy in this period. The sea-based plane's advantage against surface ships was enhanced by the development of satellite ocean-surveillance systems during the 1960s, to the point where today the daily positions of all major surface ships are known by the military. In the 1950s and 1960s, the carrier also played a role, albeit a more limited one, in wide-area search antisubmarine warfare. At that time, the United States had

---

FIG. 34

In this figure the U.S. and the Soviet Navies are compared, for all ships above 1,000 tons, the cut-off usual for ocean-going ships. The strong U.S. power projection forces, carriers, amphibious ships, and fast underway replenishment ships, are without any real counterpart from the Soviet side. But the U.S.S.R. holds the lead in anti-ship subs, and a still bigger one in old-style diesel subs. (The text explains why we cannot ascribe much anti-sub capability to the many Soviet nuclear subs.) The Soviet naval forte is anti-ship warfare, evidently with good reason. Again there is little sense in comparing forces simply by counting; more insight is needed. The U.S. fleet is overall almost twice the weight of the Soviet fleet, although it has rather fewer ocean-going ships.

Ships and their full-load displacement were enumerated and summed from the Couhat reference above. A quite close overall check on our two total fleet displacements is found in DEFENSE SECRETARY FY 1978, p. 90. See also *The New York Times,* March 7, 1977.

not only 15 carriers with attack aircraft, but also about a dozen smaller carriers equipped exclusively with antisubmarine patrol aircraft, designed to detect and destroy surface running diesel-powered submarines.

During this period, the U.S.S.R. built a small fleet of oceangoing warships, for which it had no pressing need (less than 100, compared with over 500 World War II-built surface ships in the U.S. Navy). It also built a very large number of diesel-powered submarines (about 300—far more than any other country), supplemented by a small but growing number of nuclear-powered types.

By the early 1970s, the Soviet diesel-powered submarines of the 1950s were being substantially replaced by smaller numbers (moving toward a sustained level of 160) of quieter, deeper running, long-range, nuclear-powered antiship submarines. This change was paralleled on the U.S. side by the phasing out of the last 10 antisubmarine aircraft carriers and the beginning of a buildup in the number of nuclear-powered, hunter-killer (antisubmarine) submarines, which would be more effective in searching out the Soviet nuclear subs. In this case, developments in submarine technology effectively ended the usefulness of the carrier-based aircraft in antisubmarine warfare, except as a small operation (10 to 20 percent of the carrier's aircraft) for self-defense.

At the same time, Soviet ships and submarines under construction in the late 1960s and early 1970s were equipped with new model antiship cruise missiles, which were more accurate and more reliable than the first generation. The successful use of antiship missiles first by Egypt and later by Israel against ships 10 to 20 miles away reinforced the judgment that major warships are becoming increasingly vulnerable to damaging attacks by ship-to-ship missiles, even those located on small vessels. This has presented the first significant nonnuclear surface warfare threat to the big U.S. carriers and their escort ships. If not attacked first by the carrier-attack aircraft, Soviet cruise-missile-equipped surface ships and submarines, particularly those with the newer, more accurate missiles, have potential to disable the carriers or, at the least, interrupt their aircraft operations—their *raison d'être*—for shorter or longer periods. Then, carrier aircraft unable to operate, the Soviet fleet would have an advantage in the range at which it could mount an attack on U.S. surface ships—at least until the new U.S. ship-to-ship cruise missile, Harpoon, is fully developed. Old U.S. surface ship weaponry, designed primarily for use in the defensive antiaircraft and antisubmarine roles, probably would not be effective when used in the surface-to-surface mode, against ships, out to the range of the new Soviet missiles (25 to 30 miles). Thus the central role of carriers and their aircraft in surface warfare on the high seas is slowly being impaired.

As the aircraft carrier gradually loses potential cost-effectiveness in a nonnuclear naval war with the Soviet Union, first losing its ability to project an attack in preoccupation with protecting itself, and finally unable even to defend itself, other U.S. surface combatants—the cruisers, destroyers, and frigates, which hitherto have been constrained to serve only as defensive antiaircraft and antisubmarine escorts—are slowly being equipped with effective ship-to-ship missiles. These missiles bridge the gap that has existed in offensive, antisurface attack capability between large guns of the kind that once formed the primary armament on battleships (though the largest guns mounted today are only 5 inches in diameter, as against the last battleships' 16-inch mounts) and the bombs and missiles now carried over much longer ranges, but with some vulnerability, by naval aircraft. The role of aircraft against surface ship is diminishing somewhat, and that of missile-armed surface ship is increasing.

In sum, technical developments in the weapons and platforms of naval warfare have had four main effects important for shaping naval policy. Three of these tend to constrain the operations of the U.S. Navy in comparison with the situation a decade or two ago. These are, first, that the main "power projection" forces—the carriers and amphibious landing forces—are increasingly vulnerable to a disabling attack, particularly by the Soviet Union but also, increasingly by countries with coastal defense navies which acquire ship-to-ship cruise missiles. Second, the carrier no longer dominates surface naval warfare on the high oceans—that is, warfare between U.S. and Soviet surface fleets—in the manner that it once did, but is being replaced by mirror-image cruise-missile-armed combatants. Third, the more minor of the potential U.S. naval missions—naval operations against countries other than the Soviet Union, such as port blockades, strait seizures, shows of force, or offshore bombing—will tend to have a higher cost in military confrontation, as cruise-missile technology and mine warfare techniques improve and the weaponry spreads in large numbers. There are already 60 countries with cruise-missile-armed patrol boats and other surface combatants, and the number is growing. This trend also tends to inhibit the freedom of the Soviet Union to carry out similar maneuvers around the coasts of distant countries. The fourth major development is that the improvements in missiles and other weapons (air-launched, precision-guided glide bombs, long-range wire-guided torpedoes, antiaircraft weaponry, and so on), combined with the large numbers of ships and aircraft deployed by the various countries around the enclosed seas where Soviet ports are located, particularly in the Mediterranean and the Far East, mean that in all-out surface naval engagement in and around those seas, most of the surface fleets would probably be decimated quickly. That is, a simple advantage on one side or another does not exist, nor is there any possibility of a surprise attack allowing the victor to gain a wide margin of advantage and escape relatively unscathed.

In contrast, the contest for the free use of the sea lanes on the open oceans—a contest largely waged by Soviet submarines against U.S. ships and by U.S. submarines and aircraft against Soviet submarines—remains today as it was in World War II, a gradual war of attrition. Submarine sensors have improved, as have submarine stealth and submarine armament, but there have been no decisive technological breakthroughs and none are in sight. This continues to be an area where the numbers of attacking and defending units are as much a factor as their technical capabilities.

Keeping open the sea lanes to Europe is also the only mission of the U.S. Navy which is unambiguously defensive in its immediate intention (leaving aside the matter of how forces justified for the mission are actually used, as well as the question of what sort of war in Europe they supported). This mission can be estimated to account for about half of all Navy general-purpose spending. Whether this investment is justified hangs in part on a nonnaval military matter: Is it sensible to suppose that there could actually be a prolonged war in Europe involving the Soviet Union? One may ask whether any such conflict is at all likely to start, and again whether, once started, it might not end within a matter of weeks, before convoy capability was required, either through the use of nuclear weapons or through a settlement negotiated quickly to avoid their use.

In Chapter 10 we consider plausible scenarios for a war in Europe and the rationales for a U.S. defense commitment to the area. Having reviewed the possible requirements of U.S. military forces in that region and elsewhere, we return in Chapter 11 to consider factors affecting the overall size of the naval forces and the land-combat forces which we recommend.

10

POTENTIAL CONFLICTS AND OVERSEAS COMMITMENTS

The United States has both worldwide commitments and world-wide military capabilities. These commitments are perceived as being closely linked with U.S. national security and form the broad basis for U.S. military planning. In some cases, these commitments and perceived interests are obviously very important to America, such as the security of Western Europe and Japan with their close political and economic ties to America, and the stability of the Middle East with its energy and cultural ties to the United States. In other cases, U.S. interests are rather dubious, for example, in distant third world areas. This chapter will examine U.S. interest and military commitments in five major areas of the world: Europe, China, Japan, the Middle East, and certain third world areas. The most important in terms of U.S. defense spending is Europe; this is where we will begin our analysis.

(i) Europe and the Soviet Union

Europe has been a volatile battleground of ideas, peoples, and armed forces for centuries. It has witnessed the coming and going of nations and borders, as warring factions and foreign armies have fought across its countryside. The legacy of invasions across the European continent from all directions has been continued caution, tension, and mistrust. Even in the mid-twentieth century, Europe remains the most militarized area of the globe. It is this potential region of conflict, more than any other, which drives American and Soviet military spending.

The vestiges of three decades of cold-war confrontation between the United States and the Soviet Union are still reflected in a military frontier that runs through the center of Europe. Along the

NATO & WARSAW PACT NATIONS

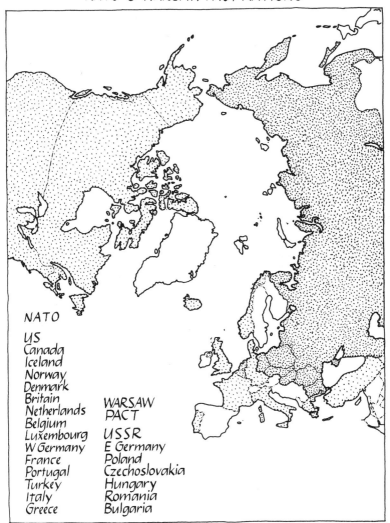

NATO

US
Canada
Iceland
Norway
Denmark
Britain
Netherlands
Belgium
Luxembourg
W Germany
France
Portugal
Turkey
Italy
Greece

WARSAW
PACT

USSR
E Germany
Poland
Czechoslovakia
Hungary
Romania
Bulgaria

FIG. 35

Armed forces of the North Atlantic Treaty Organization and of the Warsaw Pact have faced each other in Europe, across the Atlantic and the Arctic, for 30 years. NATO is strategically and economically a powerful alliance, with its 15 western side of the border stand the forces of the North Atlantic Treaty Organization (NATO), created by treaty in 1949 to assure the military security and cohesiveness of Western Europe. On the eastern side are stationed the forces of the Warsaw Treaty Organization (the "Pact") formed by treary in 1955 to defend Eastern Europe from NATO and to assure the pro-Soviet stance of the East European countries. NATO is divided into three commands—the Northern, Central, and Southern fronts. This chapter is concerned primarily with NATO Central Command—from the Baltic Sea to the Swiss Alps—where the majority of NATO, Pact, U.S., and Soviet troops are stationed (Figure 35).

THE PRESENT MILITARY BALANCE IN EUROPE

Since the ravages of World War II, Europe has recovered economically, but it has remained heavily armed, with foreign troops continually stationed on its soil through three decades of peace. NATO and the Pact expanded their forces throughout the cold war years of the 1950s and early 1960s to levels which now approach one million troops on each side. The military balance in Europe is examined in this section from several perspectives to give a full picture of these force deployments.

Three points must be recalled when analyzing the forces in Europe: (1) NATO and Pact troops are a part of much larger aggregates of worldwide strategic nuclear and conventional forces; (2) their numbers are dynamic, that is, there are frequent changes with additions and deletions of troops, tanks, aircraft, and other equipment; and (3) NATO and the Pact are different forces, both militarily and politically, with varying objectives; it is therefore not

---

industrialized members in Western Europe and North America; the Pact is composed of seven members in Eastern Europe and the Soviet Union. Three decades after the last World War in Europe, the area remains the most militarized area of the globe and the primary driving force behind American, Soviet, and world military spending.

For a brief description of both NATO and Pact alliances, see MILITARY BALANCE 1977–78. The North Atlantic Treaty, the basis of NATO, was signed in 1949; in 1952 Greece and Turkey joined, and in 1955 West Germany. The Warsaw Pact Treaty was signed in 1955; in 1968 Albania left the alliance.

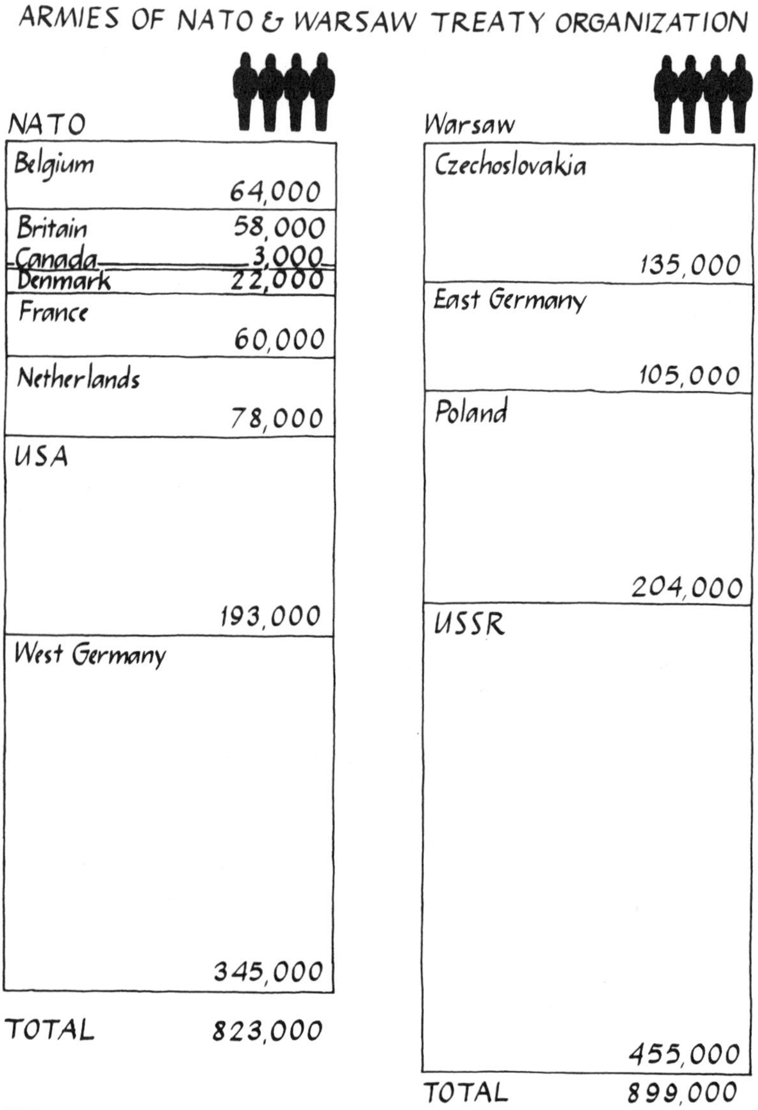

ARMIES OF NATO & WARSAW TREATY ORGANIZATION

NATO	
Belgium	64,000
Britain	58,000
Canada	3,000
Denmark	22,000
France	60,000
Netherlands	78,000
USA	193,000
West Germany	345,000
TOTAL	**823,000**

Warsaw	
Czechoslovakia	135,000
East Germany	105,000
Poland	204,000
USSR	455,000
TOTAL	**899,000**

FIG. 36

The standing combat forces of NATO and the Pact are balanced in Central Europe. Of interest is the fact that the Soviet Union contributes 50 percent of the Pact forces; if one were to discount the reliability of the East European armies, only difficult, but also wrong, to view them with mirrorlike sameness.

A first way to view the European military balance is by comparing the figures for NATO and Warsaw Pact active-duty combat and support troops in the central region of Europe; as of mid-1977, these figures are estimated as 823,000 NATO troops versus 899,000 Pact troops (Figure 36). An obvious omission from the West European forces are the 270,000 active French land forces which are not integrated in the NATO command, as distinct from the additional 60,000 French troops which are integrated in NATO and have alone been counted in our calculations.

One can note from the above figures that the Warsaw Pact slightly outnumbers NATO in Central Europe in ground forces by about 9 percent. If one were to include the Southern Command of Europe in the total—adding NATO's 540,000 troops and Pact's 395,000 troops—the ratio would be about even. The active duty ground force ratio in Central and Southern Europe is therefore quite balanced: This comparison does not take account of air forces, tactical nuclear weaponry, or available reserve troops and active forces elsewhere; these will be considered as the chapter proceeds.

A second approach to the ground force comparison in Central Europe is the NATO–Pact ratio of army divisions. The Warsaw Pact has 58 1/3 divisions versus NATO's 28 1/3 divisions, a NATO–Pact ratio of more than one to two; this seems on first glance overwhelmingly in favor of the Pact. The ratio, however, is a poor measure of the European balance. It belies the fact that there is considerable disparity between divisional sizes of NATO and Pact forces. NATO divisions include 11,000 to 18,000 troops, with both U.S. and West German ones in the 15,000 to 18,000 range. The Soviet divisions in the Pact, on the other hand, include fewer ground forces—from 11,000 to 13,000. Similarly, one may also compare divisional "slices," that is, divisions including both combat and

---

this would put the Pact numerically at a two-to-one disadvantage.

In addition to these combat troops, support and reserve troops in the Central Region and combat forces in the noncentral areas (for example, Norway and Italy, Romania and Hungary) would be brought to bear in a major engagement in Europe. However, the balanced figures indicate the unlikelihood that either side would initiate a quick attack.

MILITARY BALANCE 1977–78, and R.L. Fischer, "Defending the Central Front: The Balance of Forces," *Adelphi Papers* 127 (London: IISS, 1976).

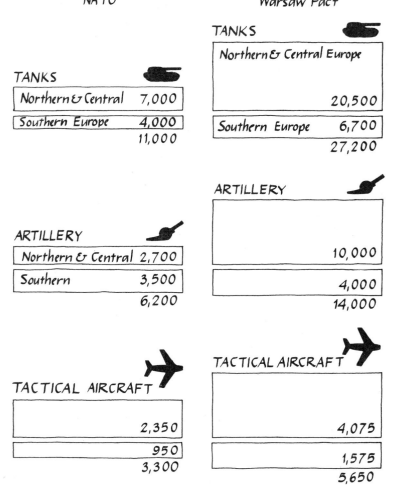

ESTIMATES OF EQUIPMENT IN EUROPE

NATO   **Warsaw Pact**

TANKS

Northern & Central Europe

                                           20,500

Southern Europe   6,700

                                           27,200

TANKS

Northern & Central	7,000
Southern Europe	4,000
	11,000

ARTILLERY

                                           10,000

                                           4,000
                                           14,000

ARTILLERY

Northern & Central	2,700
Southern	3,500
	6,200

TACTICAL AIRCRAFT

                                           4,075

                                           1,575
                                           5,650

TACTICAL AIRCRAFT

	2,350
	950
	3,300

FIG. 37

Simple number comparisons of equipment inventories can be misleading. The Pact outnumbers NATO in all categories above, yet in the aggregate serious analysts agree that the forces are near balance. Why is this so? The primary reason, among others, is the quality of equipment. NATO tanks are until now superior to those of the Pact; NATO artillery has greater lethality and higher support forces. In this case, U.S. divisional slices would be as much as 100 percent larger than Soviet and Pact slices.

A second issue which complicates any NATO–Pact divisional comparison is readiness. Some Pact divisions, particularly Polish and Czech units, are in much less than full states of readiness. As few as 50 percent of their troops may be on duty at any one time; in the event of crisis, the divisions would be filled out with reserve troops. Most Soviet divisions in Eastern Europe, however, are in a ready state although those in the Western Soviet Union are not. On the NATO side, most divisions are in a full state of readiness with the exception of British units which utilize reserves.

For these two reasons—troop numbers and readiness—one must be very cautious in comparing apparently straightforward division figures. Additional static comparisons are sometimes made between NATO and the Warsaw Pact in numbers of equipment, for example, tanks, aircraft, and artillery. Figure 37 gives estimates for these three types of military hardware.

In such quantitative comparisons, the Pact usually outnumbers NATO. These figures are the ones most often cited by proponents of increased NATO and U.S. spending on European forces. But such simple figures present a biased picture. In fact, most serious military analysts conclude that the NATO–Pact balance is just that, *balanced.* This is due to four related considerations:

1. The quality—age, readiness, and capability—of forces counts as much as does quantity of forces.
2. The quantities themselves—for example, 20,500 Pact tanks in Central Europe—can be called into question; they represent maximum authorized equipment and troop strengths of full and ready divisions. As pointed out, some Soviet and Pact divisions are in less than a full and ready state, meaning that the estimates may be overstatements of fact.
3. Numerical equipment comparisons also do not account for differences and changes in military planning and tactics; for example, NATO retains a large inventory of spare parts for tanks in order to keep as many operational as possible in battle.

---

rates of fire; NATO aircraft are longer range and carry more firepower; and NATO troops are considered more reliable than their Pact counterparts.

MILITARY BALANCE 1977–78, and MILITARY BALANCE 1976–77.

The Pact, on the other hand, cannibalizes disabled tanks in battle to keep others running; this logically necessitates a larger original inventory of tanks.

4. Some question must be raised as to the resolution and reliability of non-Soviet forces in Eastern Europe: of the 20,500 Pact tanks, an estimated 13,500 are Soviet and of the 4,075 Pact aircraft, an estimated 2,300 are actually Soviet.

A few examples here will show how incomplete—and often misleading—simple quantitative comparisons, such as those given above concerning the Central European balance, can be; these examples will help point out that NATO and Pact military forces, in spite of disparate numbers favoring one side or the other, are relatively balanced and stable.

Troops.

Military officers will readily point out that good soldiers are perhaps the primary factor in any military engagement. The American forces, who are now to a large extent volunteers, are considerably better trained and equipped than their conscripted Soviet and East European counterparts. The reliability of troops from the East European satellite nations may also be questioned if one assumes a scenario with the Pact taking the initiative. If one were to count only Soviet troops, the NATO–Pact ratio for Central Europe would change from 823,000 to 899,000 to 823,000 to 455,000, a drastic shift in favor of NATO. This is a serious issue for any Soviet officer planning forces for Central Europe.

Tactical air forces.

Combat aircraft vary widely according to mission and capability. The most offensively oriented are fighter and ground-attack planes. Of the 4,075 Pact tactical aircraft in Northern and Central Europe, only 1,350 are fighter and ground-attack planes; the rest are shorter-range defensive interceptors, reconnaissance aircraft, or light bombers. The NATO–Pact ratio in fighter and ground-attack planes for Northern and Central Europe slightly favors NATO— 1,500 to 925. If the Southern Command is included, the ratio still remains in favor of NATO—2,125 to 1,675 (Figure 38). The point to be made here, overlooked in gross numerical comparisons,

TACTICAL AIRCRAFT

NATO   Warsaw Pact

OFFENSIVE AIRCRAFT

Light bombers	150		Light bombers	175
Fighters	2125		Fighters	1675

DEFENSIVE AIRCRAFT

Interceptors	600		Interceptors	3050
Reconnaissance	425		Reconnaissance	750

FIG. 38

Any number comparisons of military forces should be explained more fully to afford more relevant conclusions. Here is an example with tactical aircraft. Although the Pact outnumbers NATO overall in aircraft, the large majority of Pact aircraft are defensive interceptors incapable of long flights and heavy loads. NATO aircraft are more offensively oriented—more capable aircraft with longer range, more maneuverability and speed, and better all-weather capability. In addition to these numbers, there are U.S. naval aircraft based on carriers, Soviet medium bombers, and American and Soviet aircraft based in the homelands; these would all be flown into a major conflict.

MILITARY BALANCE 1977–78, and MILITARY BALANCE, 1976–77.

is that many of the Pact aircraft are defensively oriented while the number of fighter and ground-attack planes favor NATO over the Pact by 30 percent. This ratio would shift further in NATO's favor if French fighter aircraft—over 400—were also included. One

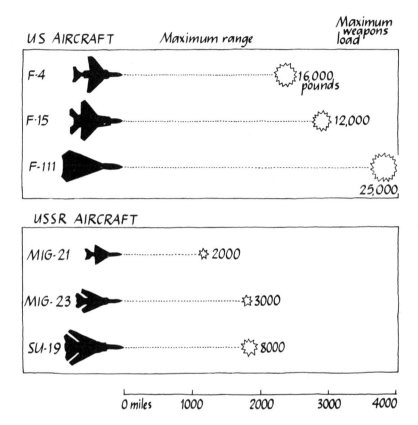

COMBAT AIRCRAFT TYPES COMPARED

US AIRCRAFT · Maximum range · Maximum weapons load

F-4 · 16,000 pounds

F-15 · 12,000

F-111 · 25,000

USSR AIRCRAFT

MIG-21 · 2000

MIG-23 · 3000

SU-19 · 8000

0 miles · 1000 · 2000 · 3000 · 4000

FIG. 39

Combat aircraft vary enormously with regard to capabilities—not to mention costs. The six aircraft here are representative of American and Soviet air forces in Europe. Notable are the large disparities in range and payload of the planes. The F-4 Phantom and the MIG-21 Fishbed are the mainstays of the forces but they are also the oldest. The F-15 Eagle and the SU-19 Fencer are the newest, both introduced in 1974, but are not yet widely deployed. The F-111, more versions of which were recently placed in Europe, is a heavy, deep interdiction fighter-bomber. The MIG-23 Flogger is a more limited supersonic fighter.

The maximum ranges are usually two or more times the typical combat mission radius. The weapons loads depend much on the mission and the variety of missiles, rockets, and bombs carried.

should also assume that, in the event of a crisis, additional reserve aircraft would be added to the picture; total aircraft available for combat in Central Europe would amount to about 6,000 for NATO and 6,800 for the Pact, still a rather close balance.

Concerning quality of aircraft, NATO planes and pilots are more capable than their Warsaw Pact counterparts. NATO pilots log considerably more flight training time and the NATO aircraft are multipurpose with much greater range, payload, maneuverability, and firepower, and better all-weather and all-day capabilities. If one compares some technical facts of the six primary U.S. and Soviet aircraft in Europe, this becomes apparent. The maximum ranges and weapons loads of the U.S. aircraft in Figure 39 are all considerably greater than those of the Soviet planes; the maximum speeds are comparable. These different capabilities allow NATO aircraft to fly, fight, and bomb deep within East European territory whereas the Pact is much more restricted to the main battle area (Figure 39). Pact deployment of defensive antiaircraft systems partially offsets this NATO advantage.

Tanks.

The NATO–Pact estimated ratio for main battle tanks in Northern and Central Europe—7,000 to 20,500—(see Figure 37) once again ignores the qualitative differences and the changing technologies of the military equipment. In field test comparisons between the primary Soviet tank—the T-62, and the primary NATO weapons—the German Leopard, the American M-60A2/3, the British Chieftain, and the French AMX-30, the NATO ones most often come out far ahead in power, speed, armor, and crew comfort. The NATO tank guns and their supporting "fire control systems" are also superior to their Soviet counterparts when one compares accuracy, range, and versatility. Soviet tanks, on the other hand, are more advanced in some categories, such as chemical and biological warfare protection and capability to ford rivers.

In judging the relative importance of tanks in ground warfare, one

MILITARY BALANCE 1977–78, and MILITARY BALANCE 1976–77; also Center for Defense Information, "Military Confrontation in Europe: Will the MBFR Talks Work?" *The Defense Monitor,* v. 4, n. 10 (December 1975); and JANE'S AIRCRAFT 1976–77.

MAIN BATTLE TANKS

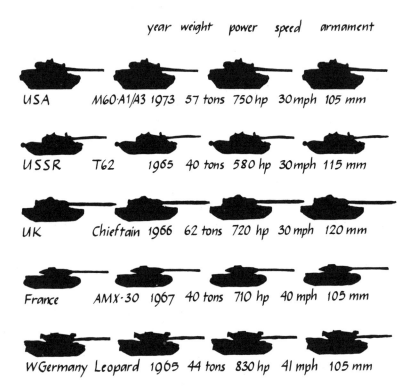

		year	weight	power	speed	armament
USA	M60·A1/A3	1973	57 tons	750 hp	30 mph	105 mm
USSR	T62	1965	40 tons	580 hp	30 mph	115 mm
UK	Chieftain	1966	62 tons	720 hp	30 mph	120 mm
France	AMX·30	1967	40 tons	710 hp	40 mph	105 mm
WGermany	Leopard	1965	44 tons	830 hp	41 mph	105 mm

FIG. 40

The five main battle tanks here are typical of present NATO and Pact equipment. NATO tanks have generally been considered far superior to Pact tanks,
although the Soviets are introducing a new T-72 which will be fully comparable
to the newer M-60s, with a better gun. The latest versions of the American M-60
are the newest tanks now deployed in number. The British Chieftain is the
heaviest, reflecting its reputation for good armor. The German Leopard is one of
the lightest with the greatest horsepower, causing it to have the highest speed.
And the French AMX presents one of the smallest silhouettes, affording it good
cover. The Soviet T-62, the tank deployed in largest number, and older Soviet
types, are simple and dependable with a heavy gun; they are relatively slow,
with limited gun range, however, and becoming old. Important characteristics
to rate the newer tanks are maneuverability, crew comfort and skill, increased
gun accuracy, firing rate, and aids to sighting.

must also consider the advantage given the defense now with highly accurate antitank missiles; a $500,000 tank can now be quite effectively knocked out with a $5,000 infantry rocket. If NATO commanders were given a choice of equipment to purchase, even with the present numerical disparity in NATO–Pact tanks, in all probability they would rather purchase 100 antitank rockets than one tank. NATO divisions are equipped with two to three times the number of antitank weapons than are their Soviet counterparts. (Figure 40 presents some technical data on main battle tanks.)

It is doubtful that NATO generals would want to go into battle with many weapons of the Pact forces. In 1975 an American general testified to Congress regarding the Soviet T–62 tank:

> I think that in almost all military technologies we lead them [the Soviets]. . . . The T–62 is really a T–54 tank [first manufactured in 1948] that has been modified a little here and a little there. . . . It has the same engine in it that the Soviets had in their tanks in WWII. There are some drawbacks to that. It isn't a powerful enough engine. . . . Our tank [the M–60] does outrange their tank. . . . I have been in the T–62 and it has a very cramped turret, and you have to be a left-handed midget because you have to load the darn thing from the wrong side of the breech. And you have to be about my size. If they run out of left-handed midgets in the Soviet Union, they are going to be in big trouble.

Both NATO and Warsaw Pact forces are currently being upgraded with new weapons systems. The Soviets are introducing additional, newer aircraft such as the MiG–23 and the Sukhoi–19, a new and more capable tank—the T–72, and a new armored personnel carrier—the BMD. The T–72 is judged about equal to present NATO tanks. On the NATO side, new tanks—the American XM–1 and a similar German one, a new armored personnel carrier—the MICV, new attack helicopters, and the F–16 fighter are being either developed or produced, and the number of FB–111 fighter-bombers has been doubled. These changes do not appear to be seriously affecting the balance.

MILITARY BALANCE 1976–77, and BROOKINGS: R.D. Lawrence and J. Record, *U.S. Force Structure in NATO: An Alternative;* also U.S. Congress, Joint Economic Committee, *Allocation of Resources in the Soviet Union and China–1977* (Washington, D.C.: U.S. GPO, 1977).

THE U.S. COMMITMENT TO NATO
AND ITS RATIONALE

The United States maintains approximately 200,000 troops in Europe. These troops are organized into four divisions, several brigades, and two armored cavalry regiments for a total of about five to six division equivalents. The divisions are all "heavy," meaning that they are made up of either armored or mechanized battalions, which have large numbers of tanks, personnel carriers, artillery, and antitank weapons. There are approximately 500 U.S. tactical aircraft in Europe, organized into seven tactical air wings of 72 aircraft each. The United States also maintains active forces in the continental United States for quick reinforcement overseas. Units which are oriented primarily toward a European contingency are the three heavy divisions which already supply one brigade each (a third of their forces) to Europe and other support forces including an armored cavalry regiment.

The U.S. commitment to NATO has two main rationales: preventing a major, conventional attack by the Warsaw Treaty Organization on Western Europe, and eliminating any need for further West German rearmament. The first broad rationale is essentially to deter aggression. Most NATO analysts, however, view the probability of such an attack as very low. With or without U.S. troops in Europe, it is doubtful the Soviet Union would want to attack Western Europe just as it is doubtful that NATO would want to attack Eastern Europe. Both sides appear to be influencing each other in direct ways—broadcasts, cultural and scientific exchanges, expanded trade and travel. They also are witnessing the rise of Eurocommunism in the West and capitalist influence in the East. Above all, perhaps, it appears doubtful that either NATO or the Pact would rationally choose to shoulder the serious political and financial problems of the other side's territory, particularly after a war.

Therefore, both politically and militarily, a war in Central Europe looks very improbable. The stakes of such a gamble are too high and the chance of winning too low to warrant such adventurism on the part of either side.

Moreover, it is not at all clear, in examining the NATO–Pact balance, that both sides have the same goals. NATO forces were built up in response to large Soviet deployments in Eastern Europe after World War II; but the Soviet forces were not in turn increased to maintain the same level of superiority. NATO and its U.S. commitment are intended to deter a conventional attack by the equally formidable Pact forces facing it in the East. This leads one to believe that the Pact forces are opposing NATO forces in only a secondary role; their primary role is to assure the political stability and friendliness of Eastern Europe. Thus, the NATO and Pact forces have very differing goals—each to defend against the other, but the Pact also to assure the U.S.S.R. of the western buffer zone, which it deems of absolute necessity to its security, between it and the Germans. Geographic and historical circumstances make it unreasonable to view both forces identically through the same pair of glasses.

The second long-range major rationale of NATO, although not spelled out as such by the United States and its allies; is to inhibit full West German rearmanent. The West German armed forces are sizable today; their active military is about half a million soldiers while their inactive reserve forces are over one million. The memories of the last World War still run deep and neither East nor West find the idea of a German superpower appealing. This point was clearly made when the West, in the 1960s, proposed sharing NATO nuclear weapons in a "Multilateral Force." The proposal was strongly opposed by many countries, including the U.S.S.R. and by some West Germans and Americans themselves, and was finally dropped.

A wide range of other reasons—military, political, and economic—are commonly used to defend a U.S. military commitment to Europe. They raise such issues as keeping Europe and NATO cohesive, inhibiting the area from developing nuclear weapons, and maintaining a strong American commitment to an area politically and economically tied to the United States. Some of these reasons for a U.S. commitment to the defense of Western Europe may be valid; some not. They cannot be looked upon, however, in a totally positive light. There are an equal number of very real and potential problems associated with U.S. troops in Europe.

These problems involve perceptions of continued U.S. hegemony in NATO and of foreign occupation of Europe by a power 3,000 miles away and 30 years after war has ceased. U.S. forces in Europe also involve heavy budgetary costs for the United States; spending estimates range anywhere from $20 to $50 billion annually for the U.S. commitment. It is therefore argued that European allies should be bearing more of the NATO burden which is directly in their backyard.

The real question remains: Given the present balanced standoff in Central Europe, and given the improbability of attack by either side, are U.S. forces still needed in Europe? Can they be restructured and reduced so as to assure Europe of a continued U.S. commitment while at the same time reducing costs and enhancing American and European security and relations? The answer is an unqualified yes.

RECOMMENDATIONS FOR U.S. NATO FORCES

A credible U.S. commitment to NATO and Western Europe does not necessarily dictate maintaining 200,000 American troops in Europe. We recommend withdrawing a U.S. division slice of combat and support troops back to the United States now—approximately 40,000 troops or 20 percent of U.S. forces in Europe—leaving U.S. forces at about 160,000, or 80 percent, for the present time.

We recognize the political sensitivity of the NATO–Pact balance based on 30 years of postwar confrontation and commitment. At the same time, the military balance in Europe is not so delicate as to make troop withdrawals difficult. Modern, emerging technologies and tactics have added much strength to the defense in almost any conflict; this was vividly illustrated in the 1973 Middle East war when defensive antitank weaponry had important impact on battlefield outcomes. It is still necessary to maintain sufficient force to deter the Warsaw Pact standing force, although a rational attack by either side has been shown to be quite improbable. A more likely scenario would be one of escalating steps of involvement. Similarly, the U.S. commitment cannot be slashed so as to undermine the perceived security of West Germany and thereby ignite full German rearmament. But steps must be taken to reduce the armed confrontation which has existed far too long in Europe.

Our judgment that a one-division withdrawal will not affect the Central European balance is based partly on military criteria. It has been shown historically that to be assured of success in warfighting an offensive force has needed at least a local three-to-one or 300 percent advantage over the defensive force. Along a lengthy front, it is estimated that an offensive force would need minimally a 25 to 50 percent advantage. With new technologies offering additional might to the defensive side, these ratios have probably increased considerably. Such offensive-defensive ratios may be measured in many ways, but in contemporary military-force planning there are four indices commonly used by military analysts to assess ratios in conventional balances. They are: so-called troop ratios, combat troop ratios, division frontage figures, and firepower scores. These may be applied to the Central European balance in order to analyze the consequences of any force increases or reductions. As long as the final NATO–Pact ratio is above 1 to 1.5, NATO will retain an adequate conventional deterrent. Above and beyond this conventional deterrent is an overly large tactical nuclear deterrent of 7,000 nuclear bombs.

The first index of measurement—the troop ratio—compares overall figures for active duty military forces on the ground. In Europe, the NATO–Pact ratio is 823,000 to 899,000 troops; after a 40,000 U.S. troop withdrawal, this changes to 783,000 to 899,000 or less than a 15 percent difference in troops. This is obviously far below the 50 percent or more difference necessary to give the Pact an unquestionable advantage in ground forces. The second index—the combat troop ratio—compares only combat troops after support troops are factored out. The infantryman would be counted in this analysis; the clerk-typist would not. The NATO–Pact combat ratio after the one-division withdrawal is approximately 431,000 to 539,000, or less than a 25 percent difference. The third index—division frontage—involves measuring the number of kilometers of battle line frontage per division. For the 900-kilometer front in Central Europe, the remaining 27 1/3 divisions after a one-division withdrawal would defend an average of 33 kilometers each; this figure is seen by U.S. military analysts as a more than adequate deterrent. The last measure of relative conventional forces is firepower scores, cumulative numbers assigned to army divisions based on how much firepower they and their weaponry can bring to bear during a limited time period. The NATO–Pact firepower ratio, after a U.S. division withdrawal, is 647,000 to 817,000, or about a 25 percent difference.

These four NATO–Pact comparisons show that, even with a 20 percent reduction of U.S. ground forces in Europe, the military balance is not seriously affected. On the contrary, there is enough margin in the ratios to allow for reductions in the future while still maintaining a militarily stable balance. A one division withdrawal of U.S. ground forces from Europe will not endanger NATO, will very likely help to lessen tensions in such a heavily armed area of world confrontation, and will act as a catalyst to the long-stagnated, multilateral effort—the Mutual and Balanced Force Reduction Talks—to reduce forces in Europe.

Air Support.

It is estimated that U.S. divisions in NATO require about 100 sorties per day per division from tactical air support in order to maintain air superiority over the Pact. Tactical air officials estimate it requires about 122 aircraft to afford 100 sorties. We therefore recommend retaining at least 500 aircraft in Europe (4 U.S. divisions × 122 = 488) to support four American divisions. This is approximately the number of U.S. aircraft in Central Europe at present. In the event of crisis, this number would be increased greatly within three to four days by reinforcements from forces based in the United States, in the fleets, and in allied territory. The United States alone could commit over 4,000 combat aircraft to the Central European region while still retaining additional aircraft in reserve. Both NATO and the Pact would expand their air forces in a confrontation, but aircraft quality and numbers in the West are more than sufficient to assure NATO of ruling the skies in any conflict.

Prepositioning.

We recognize that the European theater—or any potential area of conflict—is not a static but rather a dynamic situation. In order to judge the balance, one must not only look at present standing forces but also at the possibilities for reinforcement just before or during a conflict. If a conflict is recognized as imminent by NATO and the Warsaw Pact, both sides will in all probability seek to reinforce their commitments over the intervening weeks with additional troops and equipment before open fighting erupts.

It is estimated by the Department of Defense that airlift can transport one U.S. division and equipment to Europe in under 19 days. If that division's equipment is already in Europe, this time drops to less than 10 days. In the event of a crisis, we assume this will be considerably less. We therefore recommend that the three NATO-oriented divisions and supporting elements situated in the United States maintain their prepositioned equipment, securely stored and well maintained, in Europe. This extra complement of equipment incurs increased expenses by forcing each division to have two sets of equipment—one prepositioned in Europe and one with which to train in the United States—but the expense insures much shorter transit time for airlifting the forces to Europe in crisis. It eliminates, for example, the need in an emergency for large numbers of heavy airlift to transport 50-ton tanks and related divisional weaponry. (Airlift and sealift is discussed further in Chapter 11.)

Such a policy of prepositioning prepares the U.S. and NATO for the so-called short war scenario in Europe. It will allow the 783,000 NATO troops to be beefed up during any potential crisis by at least three U.S. divisions—120,000 troops—as well as a large number of NATO forces within a few days. These troops, with prepositioned equipment, can very quickly and easily be transported to Europe in military transport planes and commercial airliners. The additional U.S. divisions with full complements of equipment in the U.S. will take longer to transport to NATO from the U.S. by both air and sea transport, but they should certainly be in Europe in not much more than one month.

The Soviet Union is also able to reinforce its troops in Eastern Europe by moving troops from Central Russia, the Sino–Soviet border, or by activating reserves. It need not be concerned, as is the United States, with transport over 3,000 miles of ocean, but the Soviets nevertheless face equal or greater obstacles in transporting and organizing troop reinforcements overland. The Soviet Union does not have nearly the airlift capacity of the West, and Soviet railroad gauges do not match East European railroad gauges, necessitating transfers of troops in midstream. Furthermore, NATO was led to doubt the speed of Soviet reinforcements and planning during the 1968 Czechoslovak invasion by Soviet troops. The Soviets took months to deploy their forces, small relative to any hypothetical attack against NATO, before invading the East European satellite. Whatever the method, Soviet reinforcements would involve probably more than a month rather than days and would allow adequate time for NATO to react.

Additional recommendations.

A more unified NATO command structure is required. As the command structure now stands, it is too limited by national policies and interests. It seems unacceptable for the United States to maintain large forces in NATO Central on the supposition of a Soviet assault, yet have a command structure unsuited for war.

North-south mobility is also needed for NATO forces. Although the majority of NATO forces are oriented toward the center region of Europe, a Soviet attack would probably come across the north German plain.

Third, the U.S. and NATO allies must make a concerted effort at developing and procuring identical or similar weapons systems. Equipment diversity can have positive aspects; but a fighting force loses much of its punch if the ammunition does not fit the cannons, if fuel nozzles do not fit into fuel tanks, or if equipment does not fit into airlift transport. NATO countries field, for example, 23 different aircraft, 22 antitank weapons, and 7 main battle tanks. The procurement of one recent aircraft, the F-16, and the planned uniformity of new major tank parts, for example, guns, firing systems, and tracks, are all positive steps toward standardization and elimination of redundancy. This should be continued.

Finally, tactical nuclear weapons constitute a highly dangerous force in Europe. The 7,000 nuclear warheads of varying types and sizes in Europe have few, if any, rational grounds on which to base their continued maintenance. They are twice the number deployed by the Soviet Union and would be devastating for Europe if ever utilized. They must be withdrawn from Europe and dismantled; our discussion of this is outlined in Chapter 12.

We have sought to show in this chapter that the military forces in Central Europe are balanced but yet not particularly delicate. Both NATO and Warsaw Pact forces may be increased or decreased separately without drastically affecting the military mutual deterrent standoff which still continues. If present force deployments are allowed to go on indefinitely, the Central European area will remain a frontier of confrontation for the foreseeable future. Our recommendations seek to scale down that confrontation. It is difficult to predict history, but European and world security can be enhanced and economic costs can be reduced if the United States will take these first steps toward considerable force reductions.

(ii) China

The governments of China and of the United States over the past generation threw away the priceless knowledge gained by the previous generations of their merchants, teachers, students, and travelers, who had explored each others' culture, history, and people. This reciprocal break in contact continues to handicap any attempt to analyze the forces and policies of modern China, to a far

THE ARMIES OF CHINA & ITS NEIGHBORS

	manpower army, air	divisions heavy	divisions light	equipment tanks	equipment artillery
China	3,600,000	12	120	9,000	16,000
USSR	2,300,000	160	8	45,000	19,000
N Korea	500,000	3	30	2,000	5,000
India	1,100,000	4	25	1,800	2,900
Vietnam	600,000		25	900	2,500
Taiwan	460,000	14	7	1,200	1,700

The China-USSR border, armies in place

	manpower army, air	divisions heavy	divisions light	equipment tanks	equipment artillery
China ½ in the north	1,800,000	6	60	4,500	8,000
USSR ¼ on Chinese border	600,000	40	~	11,000	5,000

FIG. 41

China's armed forces are large in comparison to her neighbors, but not in comparison to her own population. The forces of North Korea and Vietnam could make a strong, local defense against any attack. The Soviet Union would probably have to transfer its European armies to engage in a serious attack on China. The whole region seems configured for defense rather than attack.

MILITARY BALANCE 1977–78.

greater extent than for any other area of the world. But China is too large, powerful, and influential to plead ignorance in defense of omission. Our analysis is brief and based on only a review of the most obvious facts.

Figures 41 and 42 indicate that China's armed forces are many times those of its neighbors, India, North Korea, and Vietnam, and not probably inferior to that part of the forces of the Soviet Union stationed in Siberia, where the two countries touch. Close to their own territory the forces of China in the 1950s secured the United States retreat from North Korea, reoccupied Tibet, and defeated Indian forces on a disputed border. They have not been used externally since, but they played a major domestic role in

AIR FORCES ~ CHINA & THE USSR

✈	Interceptors & home defense	Active tactical combat aircraft	Total aircraft
China	4000	1000	5000
USSR	2600	5300	7900

FIG. 42

The large number of older and slower Chinese planes indicates that, if attacked, China would not be without daytime defense in the air. No Chinese offensive attack on the Soviet Union could be supported by the Chinese air force for long.

MILITARY BALANCE 1977-78.

helping hold the country together during and after the Red Guard movement of the mid-1960s. China's forces appear stronger for defense of the homeland than for offense abroad, in part because mobility is a particularly expensive, industrial characteristic. The Chinese air force, for example, is large in numbers, Chinese-built but of Soviet design, two aircraft generations older than present Soviet equipment. In numbers of field artillery pieces, China ranks with the superpowers.

The interplay of ideology, opportunity, and equipment is strikingly illustrated in the Chinese navy. In the 1950s the navy began a modernization drive based on the Soviet model with many submarines and some destroyers procured. In the 1960s the emphasis—and the leading commanders—shifted. Procurement now concentrated on hundreds of home-built small coastal vessels around the 100-ton size. The navy reemphasized its links with the army, and created the water equivalent of a militia, using the immense fleet of Chinese coastal fishing and trading vessels.

As to the prospective use of Chinese military forces, some predictions can be based on the makeup of the forces themselves, history, and geography. China's coastal expansion was halted 1,000 years ago by Vietnam on the south and Korea on the north. These regional powers, with independent Communist sources of supply and substantial armies of their own, are neither satellites nor enemies, but independent states, with no incentive to attack China,

THE CHINA-SOVIET BORDER

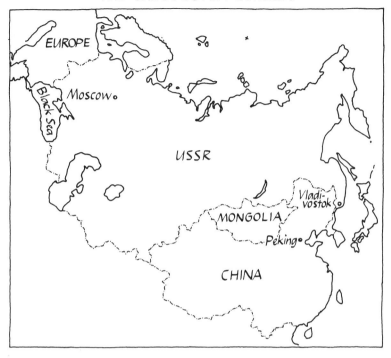

FIG. 43

Notice the distances between Moscow and the western and Black Sea borders of the U.S.S.R.; these were the great regions of World War II fighting between the U.S.S.R. and Germany. They are but a small fraction of the immense distances between European Russia and the eastern Chinese population centers. These distances and the relatively sparse rail networks of Siberia and China make supplying a major war between the U.S.S.R. and China difficult.

MILITARY BALANCE 1977–78.

and unprofitable to be attacked in return. The immense Tibetan plateau has effectively sealed China and India from land contact, and it will continue to do so. To the north, a series of desert wastes stretch over hundreds of thousands of square miles, making most of the Sino-Soviet frontier an unlikely area for major dispute.

Much has been made of the threat by 43 Soviet divisions in the east; these forces have indeed been considerably increased since the period of Sino-Soviet cooperation during the 1950s. But when factored by the length of the frontier, the major power status of China, and the semiready state of many of the Soviet forces, they are not disproportionate in size to Soviet forces along the other Soviet frontiers (Figure 43). So a Soviet attack on China seems unlikely. Nor does a Chinese attack on the Soviet Union seem possible. Peking, the northern capital of China, is almost as far from the Soviet border as it is from Canton, and the difficulties of attacking the Soviet Union must include the possibility of a Soviet counterattack which might be nuclear as well as conventional. Nonetheless, the deep-seated hostility between China and the Soviet Union cannot be totally discounted; we will return to this question.

First, however, there is the sea frontier. Will China launch some attack on Japan, or on Taiwan? There is not the slightest reason to predict an attack on Japan. The sea distances separating China and Japan are substantial, 400 to 500 miles, and China has an open-sea navy no larger than Japan's. Japan has a small, but completely modern, mobile army of its own, and Japan has the guarantee of American air and sea defense, which would be decisive against an overseas assault. Most important of all, China and Japan enjoy complementary economies and trade, and both countries would lose all and gain nothing from a war (see Figure 44).

Less is certain about Taiwan, which is regarded, after all, by both sides as an integral part of China. The history of Taiwan as the temporary refuge of the defeated previous dynasty is not new, and the actions of the present mainland regime seem cautious and patient. The island of Hainan, as large as Taiwan and only 15 miles from the mainland, was occupied immediately after mainland China fell to the Communists, but the tiny islands of Quemoy and Matsu remain to this time heavily fortified Nationalist outposts only 5 miles off the coast. Crossing 90 miles of sea to attempt to overcome the Nationalist forces on Taiwan, which though small are well-equipped and trained, would be very risky. The present American military contingent is not needed on the island, and should be withdrawn. Perhaps the passage of time, tacit negotiations with "young" Chiang, and the love of the Chinese people for China will carry Taiwan back to China eventually.

There remains the bitter quarrel between China and the Soviet Union. Our view is that the border is not the full cause of contention.

We hazard here our analysis that certain domestic strains in both countries feed the hostility. Perhaps some factions in the U.S.S.R. regarded the Chinese as potential allies; a Chinese faction once led by Army chief Lin Piao was said to have regarded the Soviet Union as allies. To control these minority groups and to block their possible appeal to the masses, a mutual blackening of the U.S.S.R. and China may have appeared desirable at some point to elements in each country.

Besides these subtle ideological strains, renewed or changed at every major succession within the Soviet or Chinese leaderships, lies the definite break over nuclear policy. Under a 1957 treaty, thousands of Soviet scientists, engineers, and technicians trained and equipped Chinese counterparts in technology, even to make nuclear weapons. In June 1959 the Soviets refused either to continue nuclear aid or to supply the Chinese with ready-to-use weapons. The Soviet exit of technicians in 1960 extended far beyond the nuclear field—all technical help ceased, all technicians left, and trade itself dropped drastically, year by year, until it almost ceased. But in October of 1964 the Chinese tested their own first nuclear weapon, and now they have probably several hundred fission and fusion warheads, deliverable by plane or missile at ranges up to 2,000 or 3,000 miles (see Figure 18). A preemptive nuclear strike against China by the Soviet Union would now present a real danger of retaliation in kind.

One recommendation does seem clear—the United States should not take the slightest step to inflame Sino-Soviet tensions, or to appear to be playing a game influenced by their quarrel. It would be folly for the United States to send arms or military supplies to China under any circumstances, or similarly to aid the Soviet Union in any military way. The United States should rather take the lead in reducing international and military tensions, so that these two countries can turn their full attention to their own domestic prosperity.

In the unlikely event of a nonnuclear Soviet-Chinese war we can predict with certainty that neither power can conquer the other. With only slightly less certainty we state that even the United States allied with one of these powers could not crush the other. The effect of nuclear conflict would be grim for both sides, though it might reflect the disparity of the nuclear arsenals. The only results of our meddling would be to widen the conflict, increase the losses, and endanger our own territory.

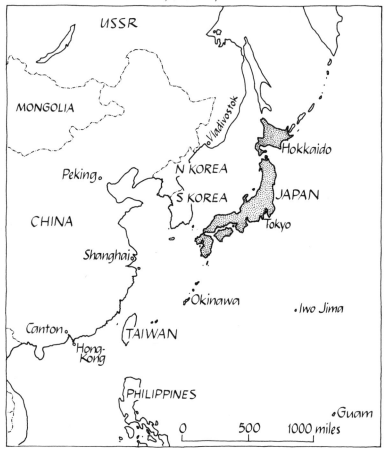

CHINA, JAPAN, GUAM

FIG. 44

The sea distances which separate Japan from China and mainland Siberia are large for a contemplated invasion, and Japan is insulated from the Koreas by a considerable water barrier. The U.S. base on Guam is positioned for supplying U.S. airpower for the defense of Japan. The American Aleutians, 1500 miles to the Northeast, provide a second line of U.S. bases for the North Pacific. U.S. bases on foreign territory are not required for the defense of Japan.

MILITARY BALANCE 1977–78.

(iii) Japan

INVASION

Japan is powerful, industrial, democratic, and like Western Europe, linked to the United States economically and by defense commitments which have been maintained for over 30 years. World War II ingrained in the Japanese public an aversion to militarism and nuclearism, which has remained to this day, embodied with U.S. participation in the very constitution of the government of Japan. In return, the United States has included Japan under the U.S. nuclear deterrent umbrella, and has guaranteed its defense against conventional attack. It is plain that Japan with its small area, dense population, and absence of national nuclear deterrence remains vulnerable indeed to nuclear threat. It is doubtful whether in fact the United States would undertake a nuclear strike by its own forces in defense of Japan, unless a nuclear attack upon U.S. territory were already certain. It seems at least as likely that Japan might choose to come to terms with a power capable of a nuclear attack on Japan which explicitly avoided involvement of the United States. The risks for all sides in such a conflict are extreme, though the likelihood of the events is quite remote.

We recommend nonetheless that these guarantees continue unchanged. If they did not continue, Japan might be more likely to remilitarize, imposing potential stresses on itself and its neighbors, without increasing the present joint U.S.–Japanese ability to repel attack.

Such an attack is extraordinarily unlikely. When, for example, the Mongols, who in 25 years had conquered mainland Asia, Russia, and the Middle East, launched the only two invasions of Japan attempted in 1,500 years, they failed. Control of sea and, nowadays, air are the essential ingredients for any possible invasion of Japan. Such a hypothetical operation must be mounted on a very large scale, as the experience of the allied invasions of North Africa, Italy, and France during World War II prove. The British Isles have been protected from invasion by a water barrier narrower than Japan's. It is worth underlining that the Japanese self-defense force must face mainly potential invaders who have managed to cross formidable sea distances of at least a hundred miles (except via Sakhalin in the North). The Japanese defense forces are far from negligible.

Even the very low eight-tenths of one percent of the gross national product of Japan spent for the military does place rich Japan among the dozen second-rank military powers, though well behind the United States and the Soviet Union. These military expenditures support armed forces of 235,000 men, including an army about one-third the size of North Korea or Vietnam, and a navy much larger than either. Japan's forces are mobile and well-equipped

Geographically, attack is plausible only from China, Korea, or the U.S.S.R. (Figure 44). China has been discussed in the previous section. It is worth repeating only that China has shown no sign of hostility to Japan, and would have to cross from 400 to 500 miles of open sea under the protection of a blue-water navy not superior to Japan's. Korea is only 100 miles from Japan, but while South Korea remains a client ally of the United States, neither reason nor ability to attack is possible. A North Korean takeover of South Korea could present a somewhat different picture; in land forces the Koreas are superior to Japan, but the lack of amphibious assault capacity or blue-water navy makes an attack impossible. Perhaps most important is the dependence of the Koreas on their respective allies for most of their military support, both financial and in actual supplies. An independent attempt by Korea to invade Japan is not conceivable.

The Soviet Union, more powerful than all other Asian countries combined, might have the theoretical capacity to attempt an invasion of Japan, if Japan were unprotected by the United States. Many reasons speak against the possibility of such an invasion.

First, the Soviet Union has a very considerable, present rival in China, with a long history of armed conflict and extreme present political antagonism. The idea of an invasion of Japan, supplied from the European Soviet power center 5,000 miles away, and passing by 2,000 miles of Chinese–Mongolian–Manchurian border, strains credulity. (See Figure 43.) Second, even without Chinese hostility, the supply line problem itself, with the inadequacy of the Soviet Pacific rail and harbor capacity, and with 200 to 400 miles of sea separating mainland Siberia from the Japanese islands, presents extraordinary obstacles. The Soviet navy has cruisers and destroyers oriented to the defense of coastal water, and only small amphibious ships and landing craft. The Soviet island of Sakhalin, it is true, is only 35 miles from Hokkaido, but the difficulties of basing a modern invasion of a large industrial nation on an undeveloped island possession are not only enormous, but slow and highly visible. Finally, the Soviet Union has had normal businesslike relations with Japan, and Japan is neither a main military nor commercial competitor.

Nonetheless, it is worth deterring even remote possibilities of invasion. The key to the defense of an island nation is air superiority. Even with total air superiority, and elements of the two largest navies in the world, U.S. General Dwight D. Eisenhower was able to cross the 100 miles of the English Channel in World War II to liberate allied France only with difficulty. Without air superiority, with a fleet not equipped for long-distance amphibious assault, a Soviet invasion seems impossible.

Air superiority for Japan's defense can and should be guaranteed by the full use of the U.S. Air Force. We discuss the strengths of the U.S. and Soviet air forces in Chapters 4, 9, 10, and 11. The United States has more and better equipment, better pilots, and experience in combat as well. The air bases in Japan from which U.S. pilots would fly are better and more numerous than those available to the Soviet Union. In case the Soviets launch all-out assaults on two fronts, Japan and Western Europe, the Soviet air force would gain no advantage, for NATO air forces would be added to the American and Japanese in opposing the Soviet. While we do not foresee the necessity of using U.S. military forces other than air, the existence of the U.S. Navy and out land forces in reserve make an already impossible invasion probably suicidal.

OTHER PRESSURES

There are other methods of pressuring Japan. A populous, resource-poor island, it is vulnerable to naval blockade especially of oil, iron ore, and food. Traditionally, a blockade has been part of full-scale war, and we have already concluded the improbability of any war from Japan's neighbors. Nevertheless a simple blockade by mine-laying, submarine, or air-and-surface attack at any point on the long sea routes which support the economy of Japan would be serious.

The issue then turns on whether shipping lanes could be kept open to the main islands for the import of some reasonable fraction of 100 million tons per year. Air support can inhibit close-in airborne mine-laying and surface attacks near the islands. The long lines from distant sources—particularly oil—would remain vulnerable. Such a blockade would of course be no swift phenomenon; it would take many months to become decisive. During these months a steady simultaneous war of attrition would take place both against the shipping, and against the blockading air or seacraft. Based on our estimates of U.S.-U.S.S.R. naval strengths, in particular as power-projected to the Pacific and Indian oceans, we think the seas could ultimately be cleared of blockade, even without direct attacks on the homeland bases which support it. This process of controlling the seas would take place much more rapidly if the blockading power were not the U.S.S.R. but some lesser power.

Of course economic activity would fall far below the present peacetime norms. Nuclear power plants and the "leakage" of some coal and oil past the blockade might allow a minimal national industry, and the food deficit would be tolerable for a long time. The possibility that emergency shipping in smaller craft might continue to the southern main islands from the Chinese coast—if assumed friendly—is one possible mitigation. The problem remains serious whatever the naval dispositions at the outset of hostilities.

Such a campaign of blockade, conducted and combated by military means, is but one point on the spectrum of possible international pressures. To take only one example from the past, the Arab oil boycott of Japan in 1973 was within weeks effective in shifting Japanese foreign policy. Indeed, Japan is only an especially vulnerable instance of the increasingly interdependent parts of the world. Europe, too, experienced economic pressures and policy changes stemming from the oil boycott.

The lesson to be learned from our analyses of Japan is that Japan is not likely to be attacked militarily. If it were, it could be defended well by the reduced military forces we recommend. But it remains an island nation, acutely sensitive even to the lesser pressures of boycott, trade tariffs, and economic up-and-down turns.

The military and logistic requirements of our analysis emphasize the need to secure the air passages from the United States to Japan via the Western Pacific. The U.S. already holds the key bases there, in particular Guam and the atoll positions en route. Prudent preparation consists in prepositioning adequate fuel reserves, navigational aids, and good communications all along that route. This is already a long-recognized naval task; the extra resources needed are minimal. Fuel reserves in Japan must be made certain. The amounts needed are not large compared to the large normal civilian use if care is taken to provide the military jet fuel necessary for the air forces. Alaska and the Aleutian Islands could in some circumstances be an alternative route to Japan. This great-circle route is shorter in total distance, but the final leg over the ocean is much longer, and passes close to the U.S.S.R. at Kamchatka and again at Sakhalin.

The present forward placement of an American Marine division and four air squadrons on Okinawa is not necessary for the defense of Japan, and is ambiguously close (500 miles) to Taiwan and to China. (See Figure 44.) It is almost as far from the American base in Guam to Okinawa (1,400 miles), as it is from Guam to the main islands of Japan (1,700 miles), and there are no islands on the flight to Okinawa. We propose returning all American forces in Okinawa and the Marine air wing in Japan to U.S. territory, to be deployed in accordance with our recommendations in Chapter 11.

In case of a surprise attack, American air reinforcements should begin arriving in Japan from Guam in hours, and from the United States in days. Within a week the entire U.S. Air Force should be available. Invasions take more than a week, and require massive, constant supplies. Supplies enough for an invading army cannot reach Japan against a superior air force.

In summary, Japan's present position in the world is peaceful and powerful, and it has no enemies. Its defensive forces, with the added protection of U.S. air forces based on U.S. territory, assure Japan's freedom from any attempt at invasion. American forces based in other countries have little relevance to the defense of Japan. We discuss them in the section on the third world later in this chapter.

(iv) Middle East

The Middle East is the oldest arena of human civilization and conflict in the world. On and off for more than 5,000 years, Egypt, Syria, Iraq, and Iran have held major, antagonistic power centers. Given the complex and shifting history of the area, it is not surprising that present conflicts are difficult to resolve or even to identify. One principle locus of actual struggle has been the bridge between Africa and Asia—the land of Israel. Another locus of potential conflict is Iran, now regaining its long-held power and influence in the world. These two are representative examples of the almost endless mesh of alliances and hostilities which mark this region (Figure 45).

The interest of the United States in the Middle East has two recent components. One component is the dominance of Middle Eastern oil in the economy of the modern world, and the necessity of

THE MIDDLE EAST

FIG. 45

Most countries in the Eastern Mediterrean and the Middle East have been armed by the U.S.A. The pressures of conflicts in which both sides were equipped with U.S. arms have shattered Cyprus and Lebanon. Iran's U.S.-supplied forces, the strongest in the region, are now prepared for action across the Persian Gulf.

U.S. naval forces in the Mediterrean are in easy reach of Israel. In addition, U.S. air power and supplies stationed in West Germany are under 1,800 miles away, within aircraft ferry or transport range.

MILITARY BALANCE 1977–78.

that oil to Western Europe and Japan, our most important economic partners and political allies. The other component stems in part from the connections binding Jews in the United States with Jews in

THE MILITARY SPECTRUM IN THE MIDDLE EAST

Manpower in thousands of men

Spending in $ billions	less than 100	100-300	300-500
5-8	Saudi Arabia		Iran
2-5	Kuwait	Israel	Egypt
1-2		Iraq Syria	
less than 1	Oman Jordan United Arab Emirates	Afghanistan	Pakistan

FIG. 46

The combination of manpower, motivation, experience, and technology over a period of time creates military power. Iran and Iraq have joined Israel, Egypt, and Syria as regional powers. Note the high spending, relative to force size, by Saudi Arabia and Kuwait, indicative of their new use of expensive, high technology and major construction projects contracted to the U.S. Army Corps of Engineers.

MILITARY BALANCE 1977–78.

Israel, and in part from the tragic extermination of European Jews during World War II, which influenced the United States and the Soviet Union to aid Israel in its creation. The fact that Israel exists in the Middle East has linked these components, and they will remain reasons for our continued involvement for at least another generation.

United States policy will have to continue insuring that the balance of arms does not go decisively or abruptly against any side in the Middle East. World opinion has demanded in many places in recent decades that the final solution of bitter rivalries not be determined by violence and death—this is no exception. But Israel does not need American military manpower, only war materiel, and the supply of war materiel does not require the stationing of large American Aircraft carrier task forces in the Mediterranean. The indigenous Israeli military forces is extremely capable by itself.

Since replacing the Soviet Union as the major arms supplies to Egypt, the United States has more potential influence in moderating the Arab–Israeli conflict than before. But it also means that the United States is in the anomalous position of arming almost all the countries on both sides of a very bitter conflict, which broke out into overt war in 1948, 1956, 1967, and 1973. The United States is selling arms for cash to some and for credit to others. While the fact that the United States is the major ally of Israel does not mean that we can dictate its policies and boundaries, it does mean we are in a position to urge moderation on Israel. We would widen the scope of our concern to include winding down the arming of both sides. The United States ought to reduce the present heavy arming of Egypt, Israel, Jordan, and Saudi Arabia. We suggest that the United States supply only weapons such as antitank and antiplane ones that tend to favor the defensive more than the offensive. Indeed, these weapons have proved effective in the two most recent Arab–Israeli wars. (See also Chapter 17.)

Should new opportunities to supply arms to Syria, Iraq, or Libya arise, economic and diplomatic aid should be tried as replacements for outright arms supply. No equilibrium position attained by war will be better, even for the victor, than a position attained by negotiations among the parties. The United States should use its influence to aid negotiations, and to persuade the Soviet Union to aid in the process. The principal contribution of the United States should be to provide diverse peacekeeping and economic aid to help implement agreements reached by the immediate participants themselves.

THE ARMING OF IRAN

In other parts of the Middle East, current United States policy is more open to criticism. Iran's oil supply is being rapidly drawn down to pay for importing Western industrialization and militarization. The rapid rise in Iranian arms expenditures and the orders for sophisticated American weaponry since 1973 (see Figure 47) is not in response to any similar rise in Soviet armaments or Soviet deployments on the Iranian borders; nor have Iranian–Soviet relations changed much in decades, except possibly for the better. An Iranian–Soviet gas pipe line has been built, and the Soviets have even supplied surface-to-air missiles and armored personnel carriers to the Iranian army. Against which neighbor then does Iran require such augmented military capacity, often of the most modern, sophisticated models available in the United States? In the east, Iran's neighbor, Afghanistan, spends less than 1 percent of Iran's expenditure on the military; Pakistan's army is large, but its military expenditures are one-tenth Iran's, and all recent Pakistani conflicts have been in India or Bengal, rather than with Iran.

To the west and south, history, opportunity, and neighbors are quite different. One neighbor is Iraq, whose rich oil fields, violently anti-Syrian and anti-Israeli regime, and recent border conflicts with Iran provide reason, excuse, and pretext for an attack by Iran. Indeed, for centuries Iraq was a territory of various Iranian emperors, of whom the present Shah of Iran is a successor. Iraq is isolated and vulnerable; while Iran has steadily and rapidly increased its armaments, Iraq has decreased its military expenditures. Both nations are beginning to enter nuclear technology, which speeds up the timetable, while increasing the stakes. To the south across the Persian Gulf—a name illustrating its history of Iranian dominance—lies a small chain of Arabian states, rich in oil, and for most of the past 1,000 years connected closely to Iran. Iran has purchased from Britain unique naval Hovercraft ideally suited for operations across the Gulf, and ordered from the United States four of the most modern Spruance-class missile-equipped destroyers. (Indeed Iran already quarters 1,000 troops in one of these states, Oman.)

Who would—who could—prevent a lightning strike by Iran against Iraq, or on these sheikdoms bordering the Persian Gulf? Iraq's allies are Algeria and Libya—and the Soviet Union. No formula for delayed explosion could be more deadly for world peace than conflict involving Iran, Iraq, and the Soviet Union. Adding to these complexities are internal long-standing tribal and social hatreds, held down by force, but capable of bursting into violent action in the distractions of modern war between rival states. We recommend that the continued supply of American weapons to Iran cease except for spare parts; we also point out the potential for U.S. involvement given the large number of American technicians servicing military equipment in Iran.

The surface forces and carrier task force we maintain in the Eastern Mediterranean are not needed to implement these recom-

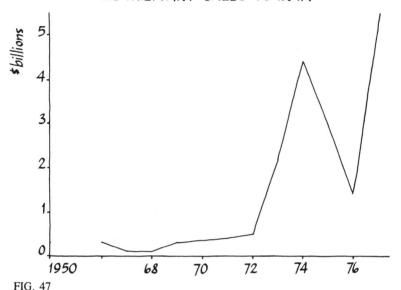

US MILITARY SALES TO IRAN

FIG. 47

The jump in arms ordered by Iran from the United States follows the jumps in oil prices during 1973. The first big peak represents the costs of the new powerful F-14 fighter and its missiles; the second represents the lower-mix light-weight F-16 fighter. There was no marked change in Iran's defense needs to require large orders of these aircraft, they are presumably available for use in offensive or interventionary operations.

U.S. Congress, House, Committee on International Relations, *United States Arms Policies in the Persian Gulf and Red Sea Areas: Past, Present and Future* (Washington, D.C.: U.S. GPO, 1977).

mended policies for the Middle East. Our presence can be effective on a much smaller military scale, as was that of the Soviet Union. Since NATO-based American forces are not far away, in time of urgency we can have air and sea control to protect communications with Israel. The American forces proposed in Chapter 11 of this book are adequate for this purpose.

(v) Other Third World Areas

American military influence spreads far beyond U.S. commitments to North America, Europe, Japan, and Israel. The spread of the political and economic influence of the United States has been accompanied by the display, and on occasion the intervention, of American military forces in many corners of the globe. This is made possible by two characteristics of current U.S. nonnuclear forces: the large size and strength of the combat elements and the long-range projection capability of the transportation elements. The size of the combat forces exceeds that needed for any probable contingency in Europe or Japan. Perhaps about half of the U.S. general-purpose forces could be pulled out of the United States, Europe, and various military bases around the globe, and inserted into a third world conflict, without disturbing the military balance vis-à-vis the Soviets to any significant degree.

The nature of the U.S. forces is such that they are particularly suited to intervention in distant parts of the world. This is particularly true of the aircraft carriers and amphibious assault ships, the sizable capability for airlift and sealift, the specially trained Marines and the more lightly equipped airborne and infantry units of the Army. All of these are suitable for interventions in third world nations, whose forces are relatively weak militarily. In addition, many current weapons (see Chapter 9) and others still under development (see Chapter 14) are primarily designed for use in such conflicts. Examples include napalm, incendiary bombs, and percussion weapons, already widely used in Vietnam.

The possession of such forces, constitutes a serious temptation. They are readily available for use whenever third world conflicts arise which can be claimed to affect U.S. interests. From time to time the United States has succumbed to this temptation, a course of action which culminated in Vietnam.

INTERVENTIONS IN THE THIRD WORLD

In the past, U.S. military intervention in third world countries has taken various forms (Figure 48). Since World War II, U.S. armed forces have been openly deployed for combat in more than a dozen nations. Except for the time of the Berlin airlift, all these engagements have been in the third world. In addition, there have been conflicts where U.S. military "advisors" played an open role, as well as other situations where it is reasonable to suspect covert action by U.S. military personnel.

The nations where the United States has been militarily involved

SOME US INTERVENTIONS IN 3RD WORLD COUNTRIES

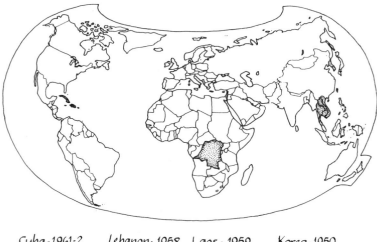

Cuba-1961-2   Lebanon-1958   Laos-1959   Korea-1950
Haiti-1962   Zaire-1964   Thailand-1968   Taiwan-1953
Dominican   Cambodia-1972   Vietnam-1963-74
  Republic-1965

FIG. 48

Since the end of World War II the U.S. has been involved militarily in over 250 conflicts worldwide, large and small. The major ones, shown here, are all in the third world and openly involved some U.S. combat forces. There are many others where U.S. military advisors have been on hand, and still others where action was covert.

BROOKINGS: B. Blechman and S. Kaplan, "Assessing U.S. Uses of Force," 1977.

are diverse, and so were the purposes of the interventions. Some were motivated by military considerations, such as the maintenance of territorial rights over portions of foreign soils. In the global competition between the superpowers for military advantage there has been a desire to deny the U.S.S.R. specific foreign bases, thus minimizing their capability to launch threats from a given area of the globe. Similarly, the maintenance and support of American forces away from the U.S. continent has required bases overseas. But an important question that should always be asked is whether the U.S. ought to have been maintaining a military presence in that region in the first place.

Another rationale for military intervention has been the containment of the spread of communism. In the geographical competition between the U.S. and U.S.S.R., this cold war concept might justify only our alliances with NATO Europe and Japan. Much more caution is required in applying a single-minded notion of containment to the third world. This rationalization has often been used to urge or to cover interventions in which primary motives included economic ones of dubious validity.

A wide range of such economic motives can be defined. The U.S. relies heavily on natural resources from many countries in the third world, and has often considereable capital investment in the industries which exploit those resources. Utilization of the cheap labor available in the third world for many labor-intensive industries has often led to capital investment by U.S. corporations. Those corporations then have a considerable interest in having the government of the third world country remain amenable to capitalist institutions, and to the repatriation of profits.

THE PROBLEMS WITH AN INTERVENTIONIST POLICY

The third world allies of the U.S.A. are often dictatorships; where there are dictatorships there tends to be resistance to the regimes. This resistance then destabilizes the country and threatens the safety of U.S. investments. Since guerrilla warfare is a common form of resistance, the United States has expended considerable effort in training indigenous armies and police forces to deal with insurgencies. But this policy still leaves open the question of what to do if the insurgents win or appear about to win. In Vietnam, the answer was an American intervention which ultimately resulted in the death of a million people and the devastation of an entire nation.

Vietnam was a shocking military and political lesson for Americans. This country learned that it was impossible to maintain an unstable regime for long by outside support. In this situation even overwhelming technological superiority does not guarantee victory against a determined adversary for whom there is significant sympathy within the country. Looking apart from the great strategic conflicts with China and the U.S.S.R., it is clear the game was not worth the candle. Some candle! (It needs to be noticed that the current government of Vietnam has seen no difficulty in seeking contracts with Western and even with U.S. firms for oil exploration off its shores—although the U.S. Congress has not concurred.)

The example of Vietnam must continually be borne in mind when evaluating the justifications for keeping U.S. forces for third world contingencies. While the cold war concept of the containment of communism may appear superficially attractive, it must be remembered that it is nearly impossible to predict the future course of events in a third world country exploding into the twentieth century. Indeed, the record of the U.S.A. in choosing which political faction to support is so poor, that the lesson to stay out has been learned. The U.S. Senate demonstrated that in its history-making decision in 1975 to keep out of Angola.

The calculus of military costs versus economic gain when U.S. inventments are at stake must not overlook the fact that it is different interests at home which pay for the costs than benefit from the gains. The International Telephone and Telegraph Corporation (ITT), for example, stood to gain from the destabilization of the Allende government in Chile with C.I.A. help, but it put up only a small part of the real cost to the U.S. taxpayer. Even more importantly the disparity between the human and economic costs of war and the gains of the investors is so huge that any justification for the U.S. action palls beside them. We need to use less outrageous sanctions for the enforcement of contracts and property rights.

SOUTH KOREA

Before discussing our policy proposals we should briefly discuss South Korea. From time to time there has been concern in defense circles about a renewed threat of invasion from the North. This case has, at least, a certain plausibility since war did break out there in 1950.

South Korea could be invaded by any combination of forces from North Korea, U.S.S.R., and China. But given the present state of Sino–Soviet relations, it is probably safe to exclude a direct Soviet invasion of South Korea (which would require supply lines through Manchuria). Chinese participation is also highly unlikely. Indeed, China is trying to gain normal relations with the United States and Japan. A war would halt the process and might provoke Japanese rearmament. Nor could China find troops to invade Korea without seriously weakening its position vis-à-vis the 43 Soviet divisions deployed along the Sino–Soviet border.

The North Koreans might act alone, but in this case there is no need for U.S. troops, even if we would be willing to send them. South Korea has a four-to-three advantage in ground troops which are well trained and equipped, a narrow front (150 miles), no serious prospect of having its flanks turned by sea, and good defensive terrain supplemented by permanent fortifications. The major concern here is not whether South Korea can feel secure vis-à-vis North Korea, but rather the reverse.

While it might be argued that internal political dissension could so demoralize the South Korean army that U.S. troops might be needed, such a proposition raises serious questions about our commitment to South Korea. If we are going to use force to defend a regime that cannot command the loyalty of its own army, then we are back to a case similar to Vietnam.

The proposed pulling back of U.S. troops (2nd Infantry Division) is a long overdue step in the right direction. We can see equally no valid reason for the remaining air force units and these too should come home. In particular we regard the presence of tactical nuclear weapons in Korea as undesirable and, as we discuss in Chapter 12, these must also be removed.

OUR POLICY PROPOSAL: NONINTERVENTION IN THE THIRD WORLD

The previous sections have indicated the frequency as well as the undesirability of U.S. military involvement in the third world. We therefore recommend some constraints on the easy continuation of previous U.S. policy. Reductions in military forces are one part of doing this; although they are an indirect way of influencing foreign policy, they are visible and unambiguous.

Bases.

Currently the United States has a series of bases on the Asian mainland and on the islands of Japan, including Okinawa and the Philippines. We recommend withdrawing our forward forces from all those points, back to U.S. bases in Alaska, Guam, and Hawaii. Facilities and supplies suitable for quick use on the invitation of the Japanese government should be retained in Japan to allow U.S. air defense against a major military threat to Japan. (For a further discussion see Chapter 10 iii.)

The situation of our Caribbean bases is more complex because of a century of U.S. hegemony. We have negotiated a return of the Canal Zone to Panama's sovereignty. A further improvement would be to give up the base at Guantanomo, Cuba. (For a more extensive discussion on U.S. bases, however, the reader is referred to Chapter 11.)

Forces.

We recommend a sharp reduction in the ready forces of the U.S.A. overseas which are not reserved for the defense of NATO. We do this because we want to minimize the potential for intervention without a mandate from the American people. Since reversals in the moderate force levels we recommend can always be made through Congress, the reductions we advocate mean no long-range risks. Our specific force proposals are made in Chapter 11 where the combined requirements of valid U.S. commitments are assessed.

11

RECOMMENDATIONS FOR THE NONNUCLEAR FORCES

In Chapter 9 we described the nature of modern war and its weapons on land, at sea, and in the air. Excluding nuclear weapons, which have not been used in warfare except in the closing days of World War II, we tried to tell what conventional war is today, and to hazard some ideas about the near future. But war is of course not merely a technical activity. It is profoundly political and geographical, not to say economic, at root. With this in mind, in Chapter 10 we set forth the issues, the opponents, and the locales where U.S. military force might plausibly be used today. In the light of our view of what the American people expect for the defense of their homeland, and for the support of those allies to whom we are now bound, we presented in the regional subsections of Chapter 10 what we hold are reasonable military commitments and military forces.

This chapter shows the total impact of the various regional recommendations when taken together. First it gives an overall quantitative account of the present U.S. military forces. Then it tells how we believe these forces should change, moving along a path toward a genuine defense. The chapter stays close to the combat elements of the forces—that is, the "teeth" of the military organization. The noncombat support activities which form the "tail"—in particular intelligence communications, research, and reserve training—are reviewed in the next part of the book. Only the forces themselves are analyzed here: The costs of the recommended combat and noncombat elements and the impact on the overall military budget are estimated in the final part of the book (see Chapter 19).

In Chapter 10 we argued that there is only one part of the world where U.S. military involvement is justified by historical bonds, political sympathies, and economic interests *combined with* (a) the presence of massed opposing military forces, which would permit a major nonnuclear war given a deterioration of political relations, and (b) the possibility of substantially increased local hostility and militarization in the event of an abrupt cancellation of an existing U.S. military commitment. This is Europe. The European war contingency is the only one for which we think it necessary to maintain substantial numbers of U.S. ground troops, to be ready to reinforce other NATO forces in case of need. Other areas which we believe would command considerable popular support for U.S. aid, if required to assure national viability, are Japan and Israel. In neither of these areas is there an existing U.S. commitment of ground forces; and we believe that no requirement for U.S. ground troops will arise over the next 10 to 20 years. U.S. commitments to the two regions—tactical air and antisubmarine warfare reinforcements in the case of Japan and military supplies and control of a supply route in the case of Israel—should, in our view, be maintained.

Within the framework of these assessments, this chapter sets out specific proposals for U.S. ground troops, tactical air forces, and naval forces. It then looks at the question of where the forces are stationed. The matter of whether or not the United States should maintain overseas military bases, and if so, where, how many, and how large, affects many other things: the need for long-range transport capability; the functions, size, and area of operation of combat ships; the degree of political support of local governments; and the potential for military involvement in local conflicts. The final section of the chapter considers the requirements for transport aircraft and ships (the "airlift" and "sealift" part of the military budget). These are needed mainly to support ground troops in Europe in the event of a prolonged nonnuclear war there.

(i) The Size and Equipment of the Combat Forces

GROUND FORCES

The basic unit of land combat in the U.S. Army and Marines is the division. A division is made up of nine to 11 battalions of various types. Alternatively, three battalions combined with a headquarters unit and some support forces can make up an intermediate-sized fighting unit, about one-third the strength of a division, called a "brigade."

As noted in Figure 26, Chapter 9, different types of division (and brigade and battalion) differ in their numbers of tanks and other heavily armored, tracked combat vehicles. The U.S. divisions equipped with many tanks are called "heavy" and the other "light."

Figure 49 sums up the equivalent divisional strength of the Army and the Marines, both active and reserve, showing where such units are deployed. Counting independent brigades at about three per division, there are twenty and one-third "division-equivalents" in all—ten heavy and ten light.

We propose to reduce the strength in heavy divisions by much less than we cut back on light divisions. The equipment and training of these armored units suit them for a major part in the defense of NATO Europe. To make perfectly plain their defensive orientation and to break the deadlock on NATO and Warsaw Pact forces which has existed for 20 years, we recommend that one and one-third division-equivalent and support (40,000 men or one-fifth of the U.S. ground forces in Europe) be withdrawn to the United States: two brigades each from the 3rd and the 8th Mechanized Divisions. This will leave the headquarters and structure for seven divisions in Europe, just as at present, though in division strength there will be only four (including the Berlin Brigade, which is light) as against five and one-third today. We also recommend an actual cut of one division equivalent which is maintained in the United States to reinforce the European troops but which does not have any component actually stationed in Europe, composed of the 5th Mechanized Division (which is at two-thirds active strength) and the 194th Armored Brigade. In addition, we propose to transfer two of the four active brigades brought back from Europe to become roundout brigades in the reserves. This will bring the active heavy division-equivalent stationed in the United States from the present five to four and two-thirds and the total number of heavy division-equivalents from 10 to eight and one-third (Figure 50).

We find no need for most of the light divisions. They are not required for the defense of island Japan, nor do we want to keep them for intervention in third world areas and in the many states where U.S. interests are *not* served by military entry. Keeping a small amphibious force for the exigencies of small-scale trouble, evacuation of civilians and the like, we propose to retain only one of the three Marine divisions. Very likely this force will become a specialized division, capable of vertical landings from the air in European conditions, and which will keep its amphibious skills as a secondary role for a few thousand men. The many other Marine

THE DIVISIONS & WHERE THEY ARE

	Far East	US Pacific	US	Europe	Reserves
1st Armored Division				1	
3rd Armored Division				1	
3rd Mechanized Infantry				1	
8th Mechanized Infantry				1	
2nd Armored Division			1	1/3	
1st Mechanized Infantry			2/3	1/3	
4th Mechanized Infantry			1	1/3	
5th Mechanized Infantry			2/3		1/3
1st Cavalry Division			1		
6th Cavalry Brigade			1/3		
194th Armored Brigade			1/3		
TOTAL heavy Army, 10					
2nd Infantry Division			1		
7th Infantry Division			2/3		1/3
9th Infantry Division			1		
24th Infantry Division			2/3		1/3
25th Infantry Division		2/3			1/3
82nd Airborne Division			1		
101st Airmobile Division			1		
Berlin Brigade				1/3	
197th Infantry Brigade			1/3		
193rd Infantry Brigade		1/3			
172nd Infantry Brigade		1/3			
1st Marine Division			1		
2nd Marine Division			1		
3rd Marine Division	1				

TOTAL, light Army, 7 1/3
    light Marine, 3

specialties—embassy guard, shipboard security, and the like—will be maintained. In the case of the Army's light divisions (seven and one-third division-equivalents), we propose to keep the more specialized units—the 82nd Airborne Division (paratroopers), the 101st Airmobile Division, which incorporates heavy use of helicopters.

In addition, we recommend that one more infantry division be retained, with one of its brigades stationed in Alaska and one maintained in the reserves. Of the three brigades in the 101st Airmobile, we recommend that one be stationed in Hawaii; some units should also be placed in Panama under the terms of the new treaty under negotiation. The 2nd Infantry Division, now being brought back from Korea, the three independent infantry brigades, and the three complete infantry divisions (each at two-thirds strength) stationed on U.S. territory could be dismantled, in bringing the Army's light divisions down to two and two-thirds. This makes an overall reduction from twenty to twelve one-third ground troop division-equivalents.

One remark is worth adding on the manpower data we have used. The public listing of the major units, divisions, and combined-armed units like brigades, is complete and explicit. Total manpower figures are also public. But we have no precise data on the important combat units which accompany the divisions, but are made up to suit needs by higher command. We have estimated their strengths, and we propose a rough figure for the size under the conditions of reduced division force.

FIG. 49

All active Army and Marine Divisions are listed with their stations as of 1978. A division normally has three brigades; the fractions show some active divisions are incomplete, rounded out by a brigade in the reserves, while others have an extra brigade overseas. Some independent brigades are not allocated to any division. Combat supplement and support units are not shown.

All light Army divisions (except the Berlin Brigade) are being pulled back to the U.S. The 2nd Infantry is beginning a slow return from Korea, though we have listed it as though it were already back in the U.S.A. The 193rd Brigade will eventually leave Panama. The 3rd Marines in Okinawa and Japan represent the only forward deployment of light forces (some are afloat with the fleets).

MANPOWER FY 1978, p. V-17ff. lists all units and their deployments.

US TROOPS & THEIR DEPLOYMENT

PRESENT	Far East	US Pacific	US	Europe & Mid-East
Divisions: Army, heavy			5	5
Army, light		$1\frac{1}{3}$	$5\frac{2}{3}$	$\frac{1}{3}$
Marine, light	1		2	
TOTAL	1	$1\frac{1}{3}$	$12\frac{2}{3}$	$5\frac{1}{3}$

PROPOSED 1980s	Far East	US Pacific	US	Europe & Mid-East
Divisions: Army, heavy			$4\frac{2}{3}$	$3\frac{2}{3}$
Army, light		1	$1\frac{2}{3}$	$\frac{1}{3}$
Marine, light			1	
TOTAL	1		$7\frac{1}{3}$	4

FIG. 50

The table gives the 1978 locations of Army and Marine divisions and our proposal to wind down these forces. Four heavy brigades from various divisions (one and one-third division equivalents) are to be pulled back from Europe, their heavy equipment is kept in place along with their headquarters structure, enabling rapid reconstruction of the division by airlift. The Marine light division in Okinawa is returned home; the light Army division still mostly in Korea is shown as already in the U.S. Almost two heavy divisions, four light Army divisions and two light Marine divisions are dismantled or placed into the Reserves to produce an active force more suited for the 1980s. The Reserves are concurrently reduced to one-third of their present manning, though most useful components are retained (see Chapter 15).

Text and MANPOWER FY 1978, V-17.

In Figure 50 we list the combat forces we recommend, to be reached by easy steps over a period of a few years. The Reserves are discussed in more detail in Chapter 15, which is devoted to reserve problems as a whole. In the figure we cite only the results of that consideration.

Finally, we emphasize that these military arguments, these forces in being, do not of themselves provide full solutions to all the questions of policy. We have indeed pruned down the ready forces to a size and stance consistent with what we have argued are the acceptable public goals of American policy today. That leaves our country with very powerful forces, still quite capable of misuse, of irresponsible or domineering power projection, especially within the world outside of NATO and Japan. Forces held for the accepted purpose of aiding our NATO allies against attack *can* be diverted to less acceptable, even to intolerable aims. Under our scheme the U.S.A. remains overall the most powerful of states. Of course it can use a small part of that strength in ways quite foreign to the design and stated intention of its forces. The intercontinental strategic bombers, the B-52s, flew in hundreds all the way from Guam, thousands of miles, to bomb the bridges and hospitals of Hanoi with high-explosive bombs, though their designers and planners thought of them as made to carry retaliatory H-bombs in a second strike against Soviet targets. Politics alone is the remedy to such miscarriages of intention.

What we think we have achieved, however, is a reduction of inappropriate but ready forces to a level where their misuse cannot be sudden, inadvertent, or go publicly unnoticed. We doubt that much more can be done as an initiative by one powerful state in the world today, without better international agreements or substantial changes in the world military climate.

TACTICAL AIR FORCES

In Chapter 9 we dissected the missions and types of aircraft which play a role in land warfare. A definitive muster of the number and capabilities of aircraft available for combat, say in the NATO region, is not as readily available as the same estimate for ground forces. Aircraft are, after all, highly mobile. In Chapter 10 we have summarized various indexes of aircraft numbers and kinds in the conventional way, to bring out a look of the NATO balance of forces. But four important conditions require statement.

1. The tactical air forces are complex. It is evident that there are scores of types and models of planes. To each model number it is commonplace to see letters affixed for new modifications: The U.S. multipurpose F–4, for example, the famous Phantom, was first tested in 1958. By now it has run through types A, B, C, D, E, F, G, and a couple more. Each time there are apt to be new engines, armament, electronics, fuel tanks, and the like. Planes are complex: For each one ready on the field to fly, more than one must be on hand. Some are always under repair, some are held to replace them, or to take care of the considerable damage rate in flight. The U.S. forces define unit equipment—the teams of airplanes that fly in squadrons— quite rigidly, to require enough aircraft, crew, and the rest to meet the numbers specified. For other forces, especially those of the Warsaw Pact, we have no such tightly detailed figures. Even the U.S.A. does not publish in the open literature any full muster of its squadrons. It does give broad overall figures—air wings, for example, or squadron numbers, organizations like divisions for the ground troops, but it does not detail the types and numbers of aircraft ready. The Joint Chiefs of Staff, for example, list the inventory of the U.S., U.S.S.R., and P.R.C. tactical aircraft. But the numbers in their report do not match the inventory numbers for the U.S.A. published in other Department of Defense documents. There is no deception; rather, accounting categories differ. For example, the Joint Chiefs of Staff report does not include spares, while the overall inventory clearly does. Obviously we have no good way to check the Soviet numbers in such fine detail.

From the figures we find, it seems a good general rule to multiply the inventory of combat aircraft (we exclude trainers and the like) by about seven-tenths to yield unit equipment, ready-to-fly with all combat needs on the line. One must of course recall that the U.S.A. has three or even four separate tactical air forces: the Air Force; the Naval tactical air wings, carrier-borne; the Marine Corps' air combat support of the Marine divisions; and the close-support U.S. Army combat aircraft, chiefly 1,000-plus attack helicopters. (We include these last in divisional equipment, and do not regard them as distinct air support.)

2. Aircraft are highly mobile. In a crisis it is clear that planes stationed in the U.S.A. or even in the Pacific could be swiflty brought to Europe, given that bases are available. It is of course important to recognize the need for fields, fuel, and manpower. About 30 airmen stand behind a typical F–4 fighter/attack plane. But it seems reasonable—and the Joint Chiefs of Staff seem to

concur—to compare total tactical air forces and not only those assigned explicitly to a given region.

3. Quality and capability of aircraft differ very widely. The most widely known combat aircraft is perhaps the MiG–21, in service with the U.S.S.R. and 20 other countries. It is a short-range, clear-weather, day interceptor, whose design was plainly first aimed at fighting off our long-range strategic bombers. Many models exist; typically this plane would take off at a 20,000-pound weight, with a modest load of weapons. A recent model of our F–4, by contrast, weighs more than twice as much, with similar or higher speed, a much greater weapons load, and much longer combat radius. Ours is a plane good for interdiction, ground attack, or air combat. Both of these planes are present in thousands in the respective forces.

How can a quantitative comparison of the tactical air forces be relied upon? The issue is far from clear. Certainly the proportion of new, powerful, all-weather aircraft strongly favors our forces. But it is clear that the expense and the smaller numbers of units has led our air force to begin building smaller planes, while the long-held lead of the U.S.A. has impelled the Soviets to begin moving toward newer, bigger and more powerful aircraft. The new MiG–23B, entering service in the early 1970s, and now present in many hundreds, is more nearly similar to an American first-line strike plane, though somewhat smaller and less able. On the other hand, direct combat in Vietnam between our F–4s and MiG–21s, while it favored the more powerful plane, showed a kill ratio much less than the cost ratio between the aircraft (about 1.5 to 1 in planes brought down). True, our planes were over enemy territory, deeply penetrating; MiG–21s are not very likely ever to go so far into such hostile skies. But some weight needs to be given to the mere provision of individual comparable aircraft; simple numbers are important even where real plane-for-plane equivalence does not exist. "The battle is not always to the strong, nor the race to the swift, but time and chance happeneth to all."

Moreover, the U.S.S.R. holds a large number of small day interceptors (mostly, but not all, even older than the MiG–21s) in their original role against our nuclear bombers, a force much larger than what we maintain for our strategic defense. Since in the present proposal we recommend steadily reducing our strategic bombers toward zero, these Soviet planes might become useful for some ground support and counterair in a land battle in Europe. Their number is large: perhaps 2,600 planes. They must be taken into account.

4. Perhaps most important, the situation is fast-changing. Five to eight years ago, the control of the air over the battlefield seemed under European conditions tantamount to victory, or at least one could not expect defeat while air superiority was on your side. But first Vietnam and then especially the intense Sinai War of 1973 has modified our expectations. In the Sinai Peninsula, inexpensive guided missiles not only brought down many expensive F-4 fighters, but even more forced a modification of figher use so that they could offer much less effective ground combat support than was foreseen. It is said that 200 F-4s accounted for only some 10 tanks of the 1,000 or more tanks destroyed in all. While one cannot be sure, the once-indispensable role of air support in ground combat between modern armies seems to be dwindling fast. Ten years ahead, manned tactical aircraft may have much less importance.

Our Proposal: Tactical Air Strength Retained

In the meantime, we draw a prudent conclusion. Since the U.S. Air Force long-range bombers will be eventually eliminated by our proposal, and since NATO land forces and politics do not seem aggressively configured for an attack on the Warsaw Pact nations, we propose to retain at present the bulk of our tactical air forces (Figure 51). The relevant naval carrier-based aircraft as well, now 65 squadrons, are much in excess of the 15 or so squadrons needed for the small remaining carrier force of our recommendation. They can be retained in active status, with gradual modification to afford combat support from land bases. This applies especially to the multipurpose F-14s and F-4s, to the reconnaissance craft, and to the close-support planes, A-7 and perhaps A-6. (Most of the more specialized carrier antisubmarine S-3s can be placed on the inactive list, in reserve status.) The Marine wings too are highly suitable for combat support, apart from a few specialized squadrons. This is an increment to U.S. Air Force strength of some 30 percent, and leaves the NATO side superior in air combat strength, or at worst in rough overall parity. These air forces hold a more favorable defensive position as the battle leaves the Eastern regions to enter the NATO lands themselves. The very numerous but short-range planes of the Warsaw powers would steadily lose relative effectiveness as they come to use fields more and more forward from their home bases.

THE DEPLOYMENT OF US AIRCRAFT

	PRESENT Far East	USPacific	US	Europe & Mid-East
Combat Aircraft: Air Force	275	– 1125 –		900
Navy	200	– 1150 –		200
Marines	150		300	
TOTAL	625	2575		1100

PROPOSED 1980s

	Far East	USPacific	US	Europe & Mid-East
Combat Aircraft Air Force		275	1125	900
Navy		200	950	200
Marines			300	150
TOTAL		475	2375	1250

helicopters are not included

FIG. 51

Here are the deployment of military aircraft in 1978 and our proposals for the 1980s. The Air Force general-purpose-forces are maintained at their present size, though aircraft currently stationed in the Far East are pulled back to U.S. territory in Guam and Hawaii, where they are still available for the support of Japan. Marine aircraft are returned from the Far East, and 150 of suitable types are restationed in Europe, to add additional air support to NATO. Most of the carrier-based aircraft of the Navy are converted to land basing. The complement from the Mediterranean carriers is kept in Europe, although their carriers are mothballed. Similarly the combat aircraft from the Pacific carriers are re-stationed in Hawaii and Guam.

FIGS. 50 and 51

Text and MANPOWER FY 1978, V-17.

We have spoken also for the defense of Japan by major dependence on tactical air. Here the naval aircraft are especially suited, but the whole force can be brought rapidly to bear. Planning for fuel and bases in Japan and for secure long ferry lines via the Pacific islands and the Marianas (or via Alaska and the Aleutians) is required over the years ahead.

Aircraft remain useful for a surprisingly long time. Still, they must be replaced at some rate. Our present aircraft are not finding replacement at rates which will keep force levels constant over an extended period, given efficient allocation of resources between maintaining old aircraft and producing new ones. This implies a continued procurement of new aircraft in some numbers over the period ahead, even though it is true that the drive for new aircraft and new armaments and electronics is one of the sharpest incentives in the technological race. On balance, we favor the maintenance of our air strength about at present level, procuring the new "low-mix" aircraft in the numbers planned, but not going at all to models beyond the ones now under procurement. The "low-mix" are F–16, and soon F–18 (the naval model intended to replace F–4, the land-based version mostly for allies); the "high-mix" F–15s continue at lower rate, while the naval F–14 can stop. (The F–4 and F–5 line should continue only insofar as it is needed to supply replacement not able to be cared for by the newer models; they are now made for export only.) The new close-support A–10 continues.

NAVAL FORCES

Since World War II, the main justification for U.S. aircraft carriers and amphibious landing ships has been to take part in major engagements against the Soviet Union, though their main use was as floating airbases off Korea and Vietnam. World War II-type amphibious operations around the coasts of Europe or in the Japanese islands during a major conventional war between the Western powers and the Soviet Union are the contingencies which are said to justify the amphibious ships. Strikes against Soviet naval forces, ports and other military and civilian targets within a few hundred miles of the Eastern and Western coasts of the U.S.S.R. represent the most important potential military mission of the carriers.

The contingencies have steadily decreased in plausibility and relevance over the period since 1965. Soviet aircraft-, ship- and submarine-based antiship weaponry presents a growing risk to the large ships. The large numbers of Soviet antiship armaments, set against the small numbers of U.S. carriers and amphibious vessels, compensates for the relatively poor technical quality of the Soviet equipment. In addition, strikes by carrier-based aircraft against land-based targets in or near the Soviet Union have lost some of their military attractiveness due to the Soviet buildup of an extensive air defense network. Finally, the perceived likelihood of a major war between the Soviet Union and the Western countries in either Europe or Asia or both has decreased markedly. As a result, the carriers and amphibious forces no longer add to the *deterrent* effect offered by our nuclear forces, ground troops, and land-based tactical air forces. This has been particularly true since around 1965, when the numbers of nuclear-delivery vehicles other than carrier-based aircraft grew very rapidly, first through the deployment of tactical nuclear weapons in Western Europe and later through the placement of multiple warheads on strategic missiles.

The only remaining issue about the role of the carriers in a major East-West war is whether such a war could be fought over an extended period of time without the introduction of nuclear weapons. Since more than half of world military spending is NATO and Warsaw Treaty expenditure on conventional military forces aimed mainly at each other, the possibility of such a war must be considered. The potential usefulness of carriers in conventional combat is much greater than in a nuclear exchange. Moreover, carriers make some contribution to deterring such a war through their usefulness in areas where staging points for land-based aircraft are few and far between. However, in any conflict with the Soviet Union, carriers would also constitute a provocative trip-wire, inviting escalation to tactical nuclear warfare. These large ships have considerable nuisance value as platforms for conventional attacks; they could at any moment become platforms for nuclear strikes and they are not located near civilians. For all three reasons they invite nuclear attack. Our conclusion is that the carriers and amphibious forces are not needed to deter a major conventional war with the Soviet Union and that they might spark nuclear exchange in the event that a conventional war broke out.

A quite different argument for maintaining the U.S. carriers and amphibious forces is to facilitate interventions in conflicts involving only much weaker military powers. Examples of such interventions are the U.S. Marine landing in Lebanon in 1958 or the offshore bombardment of Vietnam by carrier-based aircraft in the late 1960s and early 1970s. While hypothetical situations might be found where U.S. involvement of this kind would be broadly supported by the American people, the great majority of cases in which the carriers of amphibious forces might be used are likely to arouse either mixed reactions or strong sentiments against involvement. An important reason is that both forced landings and offshore bombardment suggest a weak or nonexistent popular base of support for the side the United States is aiding.

A relatively new function for U.S. carriers is to provide an imposing U.S. military presence in the Eastern Mediterranean near the scene of the Arab–Israeli conflict, and on occasion, in the Indian Ocean just outside the Persian Gulf near the lands which provide most of the oil consumed in Europe and Japan and a good fraction of the U.S. use. As long as the United States plans no commitment of ground troops to any conflict in the Middle East, neither the carriers nor the amphibious forces are needed for this purpose. U.S. submarines would be far more effective in preventing potential naval blockades of ports in the Eastern Mediterranean or Persian Gulf. In the event that some major conflict arose in which U.S. air strikes were deemed desirable, ground bases in at least one friendly country in the area—and possibly several—should be available. Similarly, U.S. carriers operating in the northwest Pacific, near Japan, could be replaced in the event of actual hostilities by operating hopping over Hawaii, Guam, and Iwo Jima to the main islands.

The main rationale for the carriers operating around the Middle East and in the Pacific is, in fact, not to provide specific military capabilities but rather to offer a highly visible and intimidating military presence as well as a hostage of considerable value, signifying U.S. political commitment and readiness for military involvement. In our view, this political mission can sometimes contribute to stability. The same is true of other, secondary military missions, such as inhibiting the closing of international straits by terrorists or Machiavellian political leaders or deterring adventurous attacks on the high seas and around naval ports. Peripheral locales—Iceland, say—might support their use even in major warfare. These contingencies offer a plausible rationale for maintaining a small number of carriers in active service—perhaps three of the present 13. (To be sure, other fast ships might do almost as well.)

If the aircraft carriers and amphibious ships were cut back substantially from their present numbers, the remaining U.S. naval forces, freed from the carrier and amphibious escort role which occupies most of them at present, would be left with a primary role of escorting military and supply convoys overseas. In essence, this means that the size of the major naval forces will be determined by a single main military requirement: keeping the sea lanes open for war-supply shipping to Europe in the event of a conventional war there lasting longer than, say, three months.

A war in Europe of many months or years duration is very unlikely. Neither the Soviet Union nor the United States and its NATO allies currently have military supplies stockpiled for more than a few weeks' engagement. The deployment of thousands of tactical nuclear weapons in and around the European theater makes it unlikely that a conflict could persist, and large territories would be at risk or lost, without nuclear weapons being introduced. Nuclear conflict would end within a few days—either through the obliteration of most targets of any value or through a cease-fire. Since the expected political and economic costs of a protracted conventional war far outweigh the benefits for all parties concerned, the most likely course of events, assuming that some crises turned into an initial open engagement, is political settlement within a matter of days or weeks.

Nevertheless, it must be admitted that if the United States were totally incapable of supporting a protracted war in Europe, the Soviet Union might gradually build up its war stocks to the point of certainty being able to outlast the West in a conventional engagement. The risks and costs of holding out would then be much lower for the Soviet Union. To prevent such an imbalance from arising, the United States should in our view continue to maintain a substantial antisubmarine warfare defense, both area and point defense, to give us a reasonable chance of getting shipborne supplies (mainly oil) through to Europe. We also recommend, however, that no further buildup of current U.S. antisubmarine forces should be made and that the United States should seek jointly with the Soviet Union to curb the submarine-antisubmarine competition for control of the sea lanes.

In sum, we make three main proposals for U.S. naval forces. First, with regard to the "power projection" forces we recommend that 10 of the 13 aircraft carriers be mothballed. The three retained in active service should be based in the United States, as training ships and for potential use as peacekeeping forces in minor contingencies. We propose to mothball all of the long-range amphibious assault ships (including the newest and largest of

A NO-GROWTH NAVY

In the active fleet:	at the end of 1976	in 1985
Aircraft carriers	13	11
Cruisers	26	24
Destroyers	69	51
Frigates	64	63

In commission for the reserves:

Cruisers	0	3
Destroyers	30	28
Frigates	0	12

FIG. 52

Ships have a useful life of around thirty years. It takes two to five years or more to build a warship, more time for larger ships, of course. Thus at any time the Navy must have many ships under construction simply to maintain its strength, even without planned increases or new types. (In late 1976 there were 106 ships in all authorized by Congress but not yet delivered.) This figure displays our estimate of the effect of stopping new construction of the main warship types to allow the fleet size to decrease by aging alone.

Once a ship is actually under construction, the equity of the taxpayers has grown to considerable size. We took that into account, and allowed new-ship completion whenever it seemed practical, ship by ship. The cut-off date for new starts was set at about 1980.

The ships used for training by the Reserves were kept almost fixed in number, by retiring overage ships entirely, placing in the Reserves the oldest ships still not overage, to reach the desired Reserve number. A few extra ships were added to the Reserves from certain classes which were distinctly older than similar ships kept active.

The ships retained, active and Reserve alike, would of course include the newest ones. Our recommendation for a surface fleet in 1985 is a little smaller than that which would be provided by attrition alone.

General information in DEFENSE SECRETARY FY 1978, p. 177ff. Individual ship lists in JANE'S and Couhat give ages and building plans. The calculation shows these intermediate results: carriers, 1 building and 3 are retired as overage. Cruisers, 5 building, 4 overage, 3 placed in Reserves as first step to retirement. Destroyers, 28 building (the assembly-line for the *Spruance* class, continued to 1980), 18 retired as overage, 28 placed in Reserves to replace the

Tarawa class), keeping a small number of amphibious coastal patrol craft (200 tons) as a coastal defense supplement to the Coast Guard.

Second, we recommend that other major surface ships—the 150 post-World War II construction cruisers, destroyers, and frigates—be retained largely as they are at the start of our program, and then decline very slowly to somewhat above 100 at the end of 10 years, mainly through natural retirements. This will still give the United States a surface navy of roughly the same size and composition as that of the Soviet Union. It provides ample surface protection for wartime convoys to Europe and offers large and well-armed ships to replace the carriers in peacetime naval visits to Europe, the Mediterranean, and the Far East. Here subs and aircraft will surely take on a larger and larger role.

Third, given these proposals, no new ship-building will be needed during the next decade (Figures 52 and 53). Thus, we recommend that a special program be established to mothball construction facilities, to provide for the retention of critical skills and knowledge, and to find alternative employment for the skilled and unskilled workers of the ship-building industry.

(ii) Operations Overseas

AIRLIFT AND SEALIFT

No other nation can rival the U.S. ability to project military power overseas, swiftly and at great distances (Figure 55). In earlier parts of this chapter, we have indicated that this capability makes it easier for the government to involve U.S. military forces in foreign conflicts even in instances where the commitment is not supported by a broad national consensus. At the same time, it is a capability justified to some by the global scope of U.S. economic interests and dependencies as well as by the far-flung network of traditional allies, from Europe to Japan.

---

28 already there in 1976, all since overage. Frigates, 11 building, none overage, but 12 of the oldest placed in Reserves. Anti-sub nuclear submarines, 16 building, 14 overage (strictly not all old, but of the smaller or one-of-a-kind types built before 1961). The net effect is given in the figure.

THE US NAVY & ITS DEPLOYMENT

PRESENT

Ships:	Far East	US Pacific	US	Europe & Mid-East
Aircraft carriers	-2-		9	2
Major gun & missile ships	-19-		127	16
Attack & patrol submarines	-15-		51	12
Amphibious warfare ships	-10-		48	5

PROPOSED 1980s

Ships:				
Aircraft carriers			3	
Major gun & missile ships		19	90	16
Attack & patrol submarines		15	43	12
Amphibious warfare ships			5	5

FIG. 53

The present and proposed deployment of the U.S. Navy are presented. The naval aircraft carriers are reduced to three, stationed on the coasts of the U.S.A. No new cruisers, destroyers, frigates, or attack submarines are started, so their numbers fall slowly as ships reach retirement age. (See Figure 52.) However, because these ships no longer have to escort carriers, effective naval strength is maintained, with a smaller force of more modern ships. Amphibious warfare ships are no longer suited for the 1980s and only two amphibious groups (10 ships) are kept for special contingencies.

Text and MANPOWER FY 1978, V-17.

Certain combat forces, already discussed, are especially designed to project power to distant areas where this country does not already have a firm foothold in popular support: that is, a lack of adequate bases. These are the aircraft carriers and the large, long-range amphibious assault carriers already discussed. Here we treat other miliary forces, which do not comprise weapons systems or direct combat capability but which can simply transport combat forces over very long ranges, as across the oceans. This section covers both air transport and sea transport capabilities.

Long-Range Airlift

Intercontinental air transport, or airlift, or ground troops and their weaponry and support equipment, is provided by two main types of Air Force aircraft, the C-5A and the C-141. About 235 C-141 jet transports comprise the mainstay of the force. These aircraft are improved versions of the familiar civil transoceanic four-engine jets, like the Boeing 707. Each C-141 is fitted out to carry 150 soldiers and all their gear, weaponry, and equipment except for the heaviest tanks and guns. The aggregate capacity of the C-141 force is nearly doubled by the Civil Reserve Air Fleet (CRAF)—some 240 long-range aircraft in civil use, about two-thirds of which are cargo aircraft or convertible to cargo configuration.

The main purpose of these long-range transport jets is to permit the rapid reinforcement and replacement of ground troops in Western Europe. They also offer the possibility of rapid, but relatively lightly armed intervention in virtually any other part of the globe. As long as the United States retains any significant capability to participate in a large conventional war in NATO Europe, however, it will be impossible totally to prevent the use of a fraction of what is rationalized for NATO in some entirely different mission. In the light of our heavy cuts in other U.S. forces, we recommend that there be no cuts in the long-range airlift forces. These force cuts, we believe, will in any case reduce the impulses to overseas adventures in other parts of the world. Moreover, we do not oppose the present plan to improve these forces by spending about $1.5 billion over a five-year period, about half this sum to increase the capacity of the existing C-141s and half to modify an additional 100 wide-bodied commercial jets for possible emergency use.

The function of the other main, long-range jet transport aircraft, the C-5A, is to carry the heaviest pieces of equipment for combat in

US MILITARY PERSONNEL OUTSIDE THE 48 STATES

Alaska	22,000	UK	20,000	Africa & Middle East		5,100
Hawaii	43,000	W Germany	220,000			
Canal Zone	9,000	Italy	12,000	S Korea		40,000
Cuba & Puerto Rico	6,000	Spain & Portugal	10,500	Japan		45,000
				Guam		9,000
Rest of the Americas	5,500	Rest of Europe	16,000	Philippines		15,000
				Taiwan		1,700
AFLOAT				Rest of E Asia & Pacific		1,100
European waters		28,000				
E Asia & Pacific		28,000		TOTAL		540,000
Elsewhere		1,100				

FIG. 54

About a quarter of the U.S. military are now outside the forty-eight contiguous states of the Union. The largest contingent is in West Germany, and more than half are in Europe, in support of NATO. The next largest number is in the Far East (to be reduced by the return home of ground troops from Korea, which began during 1978). U.S. territory and the high seas hold most of the rest. The United States operates big overseas bases around Manila in the Philippines and in Okinawa (Japan), besides the NATO lands.

At least a few men are stationed in 101 countries; 39 of the men are actually in

Europe—the biggest tanks, their auxiliaries, and some artillery—which weigh from 50 to 70 tons each and for which there is no extensive modern road network anywhere in the world outside Europe. In theory, the 68 C–5As can each carry 100-ton payloads across the Atlantic at jet speed. Despite their enormous size and great capacity, they do not have extreme range: With maximum fuel they reach about 7,500 miles, only a few hundred miles more than the C–141. The C–5As have had a long history of problems, including evidence of serious structural failure in the wings. The Defense Department is currently proposing to spend about $1 billion to strengthen the wings, arguing that this would permit the planes to last out the century.

Full prepositioning of heavy equipment in Europe, as we recommend above, reduces the need for heavy air transport, but does not eliminate it altogether. The need for combat replacement and the ability to convey equipment safe from submarine attack remain important goals in preparation for the small possibility of a short intense conventional war. Thus, the ousize transport capability may be of value in a European or even Japanese crisis even though it adds neither range nor speed to the big C–141 force.

We propose to begin a slow refit of the C–5As, but to place them formally in a new "strategic reserve condition." They would be out of mothballs, yet not on active service. To use them would require a few days of preparation, following a special and conspicuous decision by the Secretary of Defense.

While retaining this hedge, we note that the main problem of distant logistic supply, particularly in Europe and Japan in the event of any prolonged conventional conflict, is gasoline and jet fuel for tanks, trucks, and military aircraft. In both Europe and Japan, the civil use of similar products is already far larger than possible combat use. Thus, the United States could preposition an ample supply of just the needed fuels in these areas, as we now preposition weapons, dispersed for easy access and for safety. A million tons of fuel in Japan and a few million tons in Europe—a fraction of the

---

the U.S.S.R.! Significant numbers, though smaller than those here listed, are also found in Belgium, Iran, Saudi Arabia, and Turkey.

All personnel overseas are accounted for by country in *Selected Manpower Statistics* of the Department of Defense, a table on DOD MILITARY PERSONNEL STRENGTHS: WORLDWIDE. The table is issued frequently; our data refer to 1977.

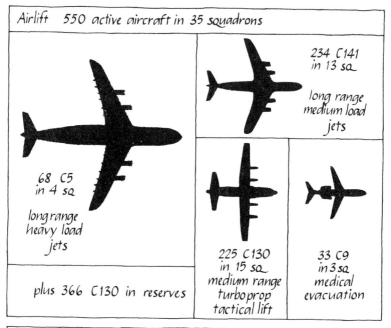

Airlift   550 active aircraft in 35 squadrons

234 C141
in 13 sq

long range
medium load
jets

68 C5
in 4 sq

long range
heavy load
jets

plus 366 C130 in reserves

225 C130
in 15 sq
medium range
turboprop
tactical lift

33 C9
in 3 sq
medical
evacuation

Sealift: unarmed ships

25 dry cargo ships
30 tankers

Operated by Military Sealift Command,
US Navy
Some Navy owned, some under charter

plus: 440 ships in various reserve categories
30 available on 10 days' notice
130 available after 60 days

FIG. 55

These unarmed aircraft have important military functions. The long-range jets
form the strategic airlift; they carry troops and equipment across the oceans,
especially in the event of an attack upon NATO. The heavy-load jets allow air
transport for the outsized fifty-ton tanks and heavy guns of the divisions when
sea transport would be too slow. The medium-load jets can carry everything
else. The smaller turboprop C-130s are tactical airlift, meant to haul troops and

total now stored there for civil uses—would offer a sizable amount for a month's full-scale combat. This, together with the air transport forces, would free our ground troops and tactical air forces of any need to use sea tanker supply for the first month or even longer in any conventional conflict. Since the costs of this approach appear relatively small, especially when set against the costs of standing naval readiness for sea convoy, we are inclined to support it.

Note that there is also a Tactical Airlift Command. About 60 percent of this is held by reserve units; in the active forces, some 240 middle-sized turbo prop aircraft (C-130s) of this command haul men and general military cargo at lesser speed over medium range, within a single theater, say Europe, but hardly across oceans. (See Chapter 9.)

Sealift

A modest force of some forty or fifty unarmed, civilian-manned ships now exists for sea transport of cargo, weapons, and fuel (Figure 55). This includes a couple of dozen moderate-sized tankers and about the same number of ships. These ships provide a peacetime light sealift capability (carrying many a GI's family auto!). Primary reliance in wartime is placed upon civil merchant ships, for their big routine transport capacity, and upon the costly naval forces discussed in earlier sections—attack submarines, patrol aircraft, and convoy escorts—for the necessary defense of

---

light-weight equipment over smaller distances within one theater, say around Western Europe. The tactical airlift is particularly well represented in the Reserves.

The sealift is of utility mainly in peacetime, servicing the forces overseas. It acts as a small nucleus for the much larger number of ships which would be required after mobilization.

Both the airlift and the sealift would be very much augmented by aircraft and ships taken from commercial service during an emergency. (See the text.)

This command also operates the specialized squadrons which carry out search and rescue for downed aircraft, five active, six Reserve.

Numbers from MANPOWER FY 1978, p. XIII-23, and JOINT CHIEFS FY 1978, pp. 94 and 95. Descriptions of planes are well done in the little *Bombers in Service* book, especially for the C-130. Couhat gives the sealift ships, and the organization and use is found in JOINT CHIEFS FY 1978, p. 96.

the sea transport lines. The peacetime sealift seens worth retention as long as the ships are economically operable.

FOREIGN BASES

The United States maintains a great many military installations on foreign soil. This form of forward projection of power, which has such an ambiguous relationship with defense, is often viewed as a clear indication of our domination of many regions of the world. A closer look at the situation is needed, and fairly complete data are available (Figure 54).

A military installation is defined, somewhat legalistically, but with much reason, as a piece of real property under the control of the Department of Defense. These extend over a very wide range in size and importance from a half acre of shore leased for the placement of some unattended navigational aid to reservations as big in area and population as a whole county (Fort Hood, Texas, or Fort Bragg, North Carolina). The total number of such installations reached almost 6,000 worldwide as of mid-1976. Of these about 70 percent were within the 50 states of the U.S.A., and 27 percent in foreign areas overseas, the rest being in U.S. territory outside the 50 states. For more detailed consideration, the chief data available represent what are called principal bases, groupings of installations with their associated properties into listed and identifiable bases. The criteria for this grouping are not explicitly given, and are indeed complex. We are led to regard the listing as broadly reflecting the real situation, though of course there may be a few surprises, kept quiet for security reasons. For instance, no principal bases are listed in Ethiopia. In the spring of 1977 the government of Ethiopia suddenly required us to close a handful of active installations there. All of the half-dozen installations which at once became visible were in fact served by under a few dozens of persons each: a medical research office, communications stations, and so on. This small but strikingly random sample gives us some confidence that the principal bases officially listed, in fact, do represent the major share of our installations overseas. No doubt that to many of these bases are assigned a fair number of small, outlying places where a few men, or only infrequent visitors, carry out some function which requires a signal beacon, or a radio receiving station, or a few standby personnel over a long time.

It is true that small installations, with few people, may still be the vehicle of covert or illicit operations, or other sorts of intrigue. And they always allow a rationalization for entry with larger forces, to assist our people on the ground. It is just as true, though, that all these circumstances can go with American civil presence or even mere ownership; it cannot be treated primarily as a matter of the armed services. Their specific interventionary risks would seem to be connected with sizable missions, in numbers of men or heavy facilities.

We show the military personnel stationed abroad and their geographical distribution in Figure 54. It is clear that the largest number of bases overseas are mainly of two sorts: The bulk of them are relatively small city or suburban quarters—holding only a small fraction of a division each—which are all that we can procure to house our troops in the highly built-up areas of West Germany (and nearby). For those, we have no recommendation; probably they should remain about as they are. We propose to withdraw U.S. forces from the large bases now operated in Korea, Japan, and the Philippines. (Very likely a few installations—but no principal bases—would remain as intelligence or administrative centers, mainly in Japan.) The big bases in the Philippines (Subic Bay and Clark Field) are now the main forward means of naval and air power projection toward Asia; their closing will limit us there strongly, to operations from bases lent us with clear local support or to quick, limited incursions based on the more distant U.S. territories in the Mariana Islands or Alaska. Our base in Cuba should be closed, and the bases in Spain as well. The airlift base in the Azores is of marginal aid to a fleet and airlift which must depend on NATO bases for operations in the Mediterranean. We foresee its retention. The bases in Australia, Iceland, and Morocco appear to be justifiable for strategic communications and NATO support. Surface communications relay stations will become steadily less useful, except as hedges against the destruction of satellites. The shift to satellites was responsible for the reduction of our large station in Asmara, Ethiopia, which within a few years declined from a staff of 3,000 or more to a few dozen men before the new government of Ethiopia asked for it to be closed. (Of course, such communication stations can serve, more or less covertly, as points for intelligence sensing of various kinds, including intercepts of messages and aids to navigation of air or seaborne craft.)

The remote Indian Ocean island base being built at Diego Garcia represents entry into a new theater; as we have argued in Chapter 10, it ought to be closed. We ought to depend on the powers whose shores lie there to support the safeguarding of that important seaway.

Overall, we would expect to retain the bulk of the European troop quarters, but with fewer elsewhere, a saving in costs, and a genuine and considerable reduction in our military presence—and threat— especially in East Asia. There would be no serious loss in our capabilities in defense of our proper interests or our chief allies.

The modest if genuine change this represents is the index of the one situation in which the visible posture of our military is in fact more nearly defensive than it appears in the intrusive image cast by those grand aggregate totals of hundreds of principal bases—and 1,500 installations—in foreign lands.

12

TACTICAL NUCLEAR WEAPONS

Tactical nuclear weapons or theater nuclear forces are those nuclear weapons deployed for use against targets in local theaters of combat, as in Korea or in France. They are intended to complement conventional forces on the battlefields and to deter any enemy from initiating a nuclear war. (Strategic nuclear weapons, as pointed out in Chapters 6 and 7, are intended for use primarily against strategic targets in the homeland of an opponent, for example, the Soviet Union.)

Theater nuclear forces are extremely important in arms control and defense planning because of their large numbers and their enormous potential for death and destruction. Of the more than 30,000 nuclear weapons deployed worldwide by the United States today, over 22,000 are considered tactical in nature. The general deployment of U.S. tactical nuclear weapons is illustrated in Figure 56.

These weapons vary in size, range, and delivery system. Tactical nuclear weapons include short- and long-range surface-to-surface missiles with ranges from 25 to 450 miles and yields from 1 to 400 kilotons. They also include surface-to-air and air-to-surface missiles, dual-capable artillery for firing either nuclear- or conventional-armed shells, atomic demolition mines, and antisubmarine depth charges and torpedoes. These systems too have equally variable ranges and yields. A representative list of tactical nuclear weaponry is given in Figure 57.

The terms "tactical" or "theater" do not therefore necessarily denote small yield and destructive potential or short range. Their ranges are short only relative to the several-thousand-mile ranges of strategic missiles; they are still capable both of striking troop concentrations on the nearby battlefield and of reaching targets in the rear zones for purposes of interdiction. Some of these weapons

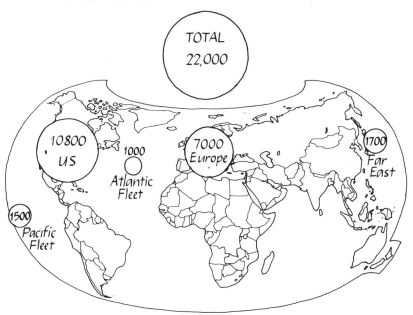

US TACTICAL NUCLEAR WEAPON DEPLOYMENT

TOTAL
22,000

10800
US

1000
Atlantic
Fleet

7000
Europe

1700
Far
East

1500
Pacific
Fleet

FIG. 56

The United States deploys over 22,000 tactical nuclear weapons around the world, most larger in yield than the two atomic bombs dropped on Japan in 1945. Nearly 50 percent of these are stored in the United States. Thirty percent are deployed with U.S. troops in NATO, while the remaining several thousand are on U.S. ships and in Korea.

Center for Defense Information, "30,000 U.S. Nuclear Weapons," *The Defense Monitor,* v. 4, n. 2 (Feb. 1975).

are 25 times larger in yield than the Hiroshima and Nagasaki bombs which killed 100,000 people each. The combined explosive capability of the tactical nuclear forces is equal in energy to about 500 megatons of TNT, 250 times the total amount of explosives dropped in World War II, and 50 times that dropped throughout the Vietnam War. Some tactical nuclear weapons are ten or more times as powerful as a present-day strategic warhead.

Tactical nuclear weapons were first introduced into the U.S. inventory in the 1950s. They were subsequently deployed in

Europe to offset a then-perceived imbalance in NATO–Warsaw Pact conventional forces. As former Secretary of Defense Rumsfeld stated in the 1977 annual defense report, tactical nuclear weapons were justified by the Defense Department on the grounds that they contributed "superior firepower to compensate for what was considered an unfavorable conventional balance." In the last 20 years tactical nuclear weapons have come to be accepted, along with strategic and conventional forces, "for defense of and defense against aggression in the Pacific and European theaters and to control escalation should deterrence fail." Thus tactical nuclear weapons are seen by some planners as enhancing both U.S. defense and the U.S. commitment to Europe and Asia. They are also seen as possible substitutes for U.S. troops, as some part of the nuclear deterrent, and as a means for providing a nuclear war option at smaller stakes than all-out strategic nuclear exchange (and on someone else's territory).

We believe these rationales do not provide sufficient justification for maintaining such a large widespread tactical nuclear weapons inventory. We arrive at our conclusion for a number of reasons.

1. There is the obvious danger of conflict escalation. The widespread deployment of tactical nuclear weapons in such areas of confrontation as Europe or Korea would only serve to raise a local conventional conflict to a nuclear one if war breaks out. With 7,000 U.S. tactical nuclear weapons present in Europe and 1,700 on the Korean mainland and other parts of Asia, it is easy to believe that they might be used in a local conflict. Now, these weapons are often indistinguishable from strategic nuclear weapons with regard to yield and range. Soviet military writers, moreover, reiterate that their nuclear doctrine does not differentiate between strategic and tactical nuclear warfare. The danger of escalation to worldwide nuclear conflict through use of such weapons is by no means unreal. It seems ironic that the United States deplores the possiblity of the Middle East conflict area becoming nuclear-armed, yet argues for nuclear deployments in Europe and Asia. All three areas offer dangerous potential for nuclear war.

2. Tactical nuclear weapons serve to increase the amount of possible collateral damage in a conflict area. The weapons are extremely destructive in themselves, creating servere radioactive fallout in addition to their immediate blast, shock, heat, light, and radiation effects. In fact, it is difficult to realize their effects, be it on a battlefield swarming with troops or in a city swarming with citizens. The weapons are intended for detonation in the Soviet

Union and Eastern Europe. But they tend as well to endanger Western Europe by inviting both preemptive and retaliatory nuclear strikes from the Soviet Union's 3,500 tactical nuclear weapons held in Eastern Europe and from Soviet-based nuclear systems. In NATO war games, simulations of tactical nuclear explosions have indicated that millions of European civilians would be killed. In one such simulated war game, called "Carte Blanche," it was estimated that about two million Germans alone would be killed and over three million injured, by 268 tactical nuclear detonations on German territory. There is no describing what over 7,000 detonations throughout densely settled East and West Europe would do. Use of tactical nuclear weapons in local conflicts would increase both military and civilian casualties and the damage to industry and the environment to a terrifying degree.

3. The deployment of thousands of tactical nuclear weapons presents an enormous security problem. One can easily imagine possibilities for accidents, for unauthorized use by U.S. or Allied forces (in spite of safeguards), or theft and blackmail by terrorists. Tactical nuclear weapons, including modern "mini-nukes," small and light enough to carry in a suitcase, are being handled yearly by thousands of personnel at hundreds of different sites. There are 100 "Special Ammunition Storage" sites in Europe alone. This important security problem was recognized in the 1950s when the United States withdrew the small, jeep-mounted tactical nuclear rocket, the Davy Crockett, from Europe because local commanders were fearful for the weapon's security. At present, the United States is seeking to improve the physical security of our weapons through such schemes as a double-key security system to prevent unauthorized launch, and a careful screen of the individuals who handle tactical nuclear weapons. The most powerful way to deal with the security problem of tactical nuclear weapons is to decrease sharply the number of such weapons at home, and to withdraw them from overseas.

Tactical nuclear weapons are no longer needed to redress perceived "imbalances" in conventional forces. Gross imbalances do not exist in conventional forces in the European and Asian theaters (see Chapter 10). The U.S.A. has many more tactical nuclear weapons in these areas—7,000 in Europe and 1,700 in Asia—than do its adversaries; the Soviet Union has 3,500 tactical nuclear weapons in Europe while in Asia, China has about 50 intermediate and medium-range ballistic missiles. These Chinese missiles are primarily targeted at the U.S.S.R. The U.S.A. also has

SOME TACTICAL NUCLEAR DELIVERY SYSTEMS

Artillery	Missiles	Aircraft
M109	**Honest John**	**F4**
155 mm self-propelled howitzer, smallest nuclear-capable gun. 2 kT shells. $500,000 each	Single-stage, unguided, 20 kT, battlefield support	Multi-use fighter, capable of dropping up to 8 tons weight of varied nuclear bombs
M110	**Lance**	**A6**
8 inch self-propelled howitzer. 2 kT, accurate, 13 in crew	New, accurate 1-100 kT, for battlefield support	Attack bomber, capable of dropping nuclear bombs on or behind the battlefield
⊛ nuclear ⁂ nuclear capable	**Pershing**	**F 111**
	Two-stage, accurate 60-400 kT, for battlefield or logistics attack	Heavy fighter-bomber capable of delivering 6 nuclear bombs, or short-range attack missiles
Up to 18 miles	Up to 450 miles	Up to 1500 miles

FIG. 57

There is a wide variety of tactical nuclear weapons in U.S. inventories. The eight examples above illustrate the differences in type of delivery system, in nuclear yields, and in ranges. Jet aircraft may carry several nuclear bombs and missiles around or behind the battlefield. Missiles with "dial-a-yield" warheads may be launched over a several-hundred-mile range. And nuclear-capable artillery may be fired for battlefield support. There are also other nuclear systems deployed, such as atomic demolition mines in Europe, and air-to-air missiles in the U.S.A. In Europe alone there are over 500 surface-to-surface missiles, a thousand plus artillery shells, and several thousand bombs, mines, and air-to-surface missiles, all nuclear armed.

BROOKINGS: J. Record, *U.S. Nuclear Weapons in Europe: Issues and Alternatives,* 1974; also MILITARY BALANCE 1977-78; and JANE'S AIRCRAFT 1976-77. The *Lance* missile is 20 feet long.

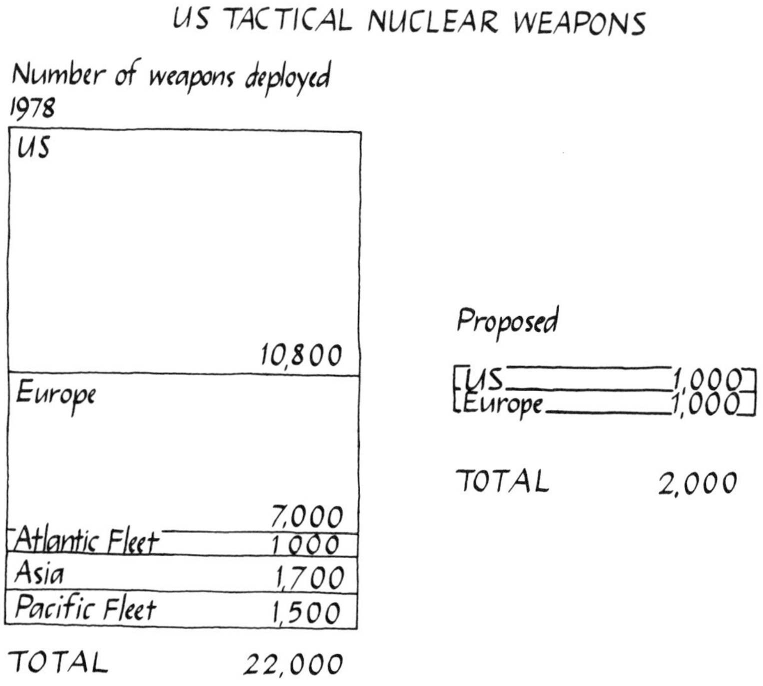

US TACTICAL NUCLEAR WEAPONS

Number of weapons deployed 1978

US	
	10,800
Europe	
	7,000
Atlantic Fleet	1,000
Asia	1,700
Pacific Fleet	1,500

TOTAL   22,000

Proposed

US	1,000
Europe	1,000

TOTAL   2,000

FIG. 58

The United States has deployed many more tactical nucleare weapons worldwide than are necessary for adequate defense of U.S. and allied forces. They appear in part a remnant of an obsolete strategy, more a danger than an aid in today's world. We propose withdrawing and dismantling 20,000 of the present 22,000 warheads. They are not needed for protection of the fleet nor of troops in Korea. One thousand are temporarily retained as a reserve in the United States and another thousand in Europe in recognition of Soviet tactical nuclear deployment there.

See note to text page 211.

strategic submarine-launched ballistic missiles targeted on certain sites in Europe, while the Soviet Union has hundreds of medium- and intermediate-range ballistic missiles presumably targeted on Western Europe. It costs over $500 million annually to maintain U.S. tactical nuclear weaponry in Europe alone. Our tactical nuclear weapons represent expensive and redundant capabilities, while offering potential for less rather than for more security. The whole purpose of tactical nuclear weapons today appears to be in doubt. They were introduced into U.S. forces to help deter a Soviet conventional first strike, and to aid in war-fighting if a major conflict broke out. At that time there were no Soviet tactical nuclear weapons opposing them. Twenty years later, the strategic nuclear forces provide more than an adequate nuclear deterrent; and the real utility of tactical nuclear weapons in theater conflicts is now in doubt, both because of their high destructiveness to the civil society and their own vulnerability to first-strikes from a Soviet tactical and medium-range nuclear force now very capable.

In light of all these arguments, it is important that the number of U.S. tactical nuclear weapons be reduced. To meet the complexity of the problem, we offer the following broad recommendations:

(1) All the 1,700 land-based tactical nuclear weapons in Asia, that is, in Korea and the Philippines, and on Guam and Midway, should be withdrawn to the United States and dismantled.

(2) The 7,000 land-based tactical nuclear weapons in Europe should be immediately reduced to 1,000. We do not recommend an immediate reduction to zero because of the sizable Soviet tactical nuclear force now in Eastern Europe. The remaining U.S. warheads should act as an incentive for mutual nuclear reductions by both NATO and the Warsaw Pact.

The number of launchers for these 1,000 warheads should not be increased above present levels. In order to delineate the intermediate threshold between theater use and full strategic nuclear warfare and to lessen the theater vulnerability to preemptive strikes, the remaining tactical nuclear weapons should consist primarily of aircraft-launched weapons of longer ranges. Such aircraft could be based off the European continent in Britain. The older, more destructive, less accurate, and less secure weaponry, such as the Pershing, Sergeant, and Honest John surface-to-surface missiles, to nuclear artillery, and the atomic demolition mines, should be withdrawn to the United States and dismantled. All tactical nuclear weapons should be removed from "quick reaction alert" aircraft. These aircraft are so staged as to respond within minutes to an attack, allowing little, if any, time to decide whether one should respond on a nuclear level or not. Quick-reaction alert aircraft with nuclear weapons are both risky and unnecessary; the vulnerability of the weapons is reduced by moving them back.

(3) All tactical nuclear weaponry on the Atlantic and Pacific fleets should be withdrawn to the United States and dismantled.

Defensive nuclear weaponry, such as surface-to-air missiles and depth charges, are claimed to be essential for defense of the fleets, but in view of the worldwide danger of breaking the nuclear barrier, restriction to conventional weaponry should be worth the risk.

(4) The 1,000 tactical nuclear weapons remaining in Europe should not be considered a final minimum. A force of over 1,000 such weapons on each side—including the several hundred submarine- and land-based ballistic missiles of the French and British intended for use in Europe—retains the capability to lay waste a large portion of the Europe continent. These weapons should be eliminated through multilateral negotations, such as the present Mutual Force Reduction talks in Vienna. The eventual goal might be a nuclear-free zone in Central Europe similar to the plan proposed 20 years ago by Adam Rapacki, the then Polish foreign minister. Even after any such cutback, U.S., British, and French sea-based strategic nuclear missiles still possess the option of targeting tactical sites in Europe from the surrounding seas.

(5) The U.S. domestic inventory of 11,000 tactical nuclear weapons should be drawn down to 1,000 immediately, with the eventual goal of zero within five to ten years.

(6) Research, development, testing, and evaluation in tactical nuclear weaponry and further production of such weapons should be immediately halted. Improved accuracy and yields have reached the point where further development is both unnecessary and destabilizing. In spite of this, both the U.S.A. and the Soviet Union are modernizing their forces. The U.S.A. has proposed introducing an "enhanced radiation" weapon—the so-called neutron bomb—into NATO forces. Such modernization is more dangerous than helpful and ought to stop. The neutron bomb risks blurring the conventional-nuclear distinction and should not be deployed. (It is an antipersonnel device of special cruelty.)

(7) Efforts at improving the security and invulnerability of the remaining tactical nuclear weaponry should be emphasized. Command and control of the weapons must remain with the central decision-making hierarchy under the President of the United States; neither local commanders nor U.S. allies should at any time have the ability on their own to order use of U.S. tactical nuclear weapons.

If the foregoing recommendations are followed, the U.S. tactical nuclear inventory will be decreased from 22,000 warheads to 1,000 in Europe and 1,000 in the United States (Figure 58). These remaining 2,000, by themselves, an incredible destructive force, must also eventually be eliminated. These measures will not only enhance security but also lead to significant savings.

The United States should also renounce the first use of tactical nuclear weapons. Although such a step would not be directly reflected in the U.S. defense budget, it would unequivocally show that the U.S.A. does not intend to escalate conventional conflict to the nuclear level.

PART IV
SUPPORT ACTIVITIES

13
INTELLIGENCE AND COMMUNICATIONS

It is no surprise for the television viewer that the flow of information, mainly—but by no means only—in electronic form, is one of the fastest growing sectors of the world economy. So it is within the domain of the American military. The 1978 program budget allots $9 billion to intelligence and communications, including $1 billion earmarked for these uses within the R&D program request. (We point out that there are hidden sums for the C.I.A., for the National Security Agency and for the National Reconnaissance Office estimated variously at $1 billion to $3 billion.) The human effort and the hardware used are of extraordinary diversity, and, on the whole, little described in the public literature. We have some limited experience with certain of these operations, but we regard the whole topic as deserving a distinct expert study. Because of its overall size and its great importance, we nevertheless devote a brief chapter to the problem, with recommendations which we label quite frankly as tentative (Figure 59).

The activities here under concern can be divided into two distinct but related functions: seeking information about the outside world, from daily weather to foreign policy of other states, and the transferring of information within our own forces, everything from the daily management of millions of people who must be fed and paid to the life and death decisions of armed conflict. The first is broadly intelligence, while the second is called command, control, and communications (often abbreviated as C3 in military jargon).

The sources of intelligence are broad and diversified, consisting of at least four general classes:

   (1) Human sources, from the openly tolerated and mutual inquiries made by military attachés in every big diplomatic mission in all capitals, through travelers' accounts, to the dramatic secret work of the clandestine agent.

INTELLIGENCE & COMMUNICATIONS BUDGET

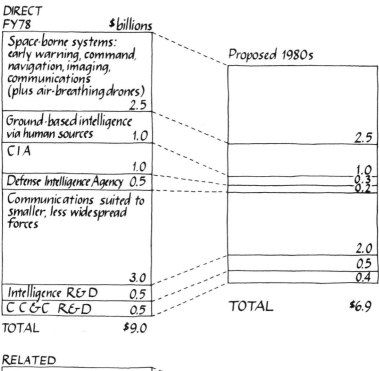

DIRECT
FY78 $billions

Space-borne systems: early warning, command, navigation, imaging, communications (plus air-breathing drones) 2.5	Proposed 1980s
Ground-based intelligence via human sources 1.0	2.5
CIA 1.0	1.0
Defense Intelligence Agency 0.5	0.3 / 0.2
Communications suited to smaller, less widespread forces 3.0	2.0 / 0.5 / 0.4
Intelligence R&D 0.5	
C C &C R&D 0.5	TOTAL $6.9
TOTAL $9.0	

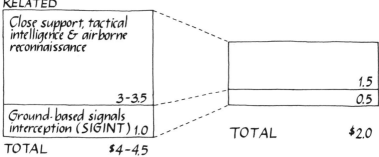

RELATED

Close support, tactical intelligence & airborne reconnaissance 3-3.5	1.5 / 0.5
Ground-based signals interception (SIGINT) 1.0	TOTAL $2.0
TOTAL $4-4.5	

FIG. 59

No portion of the budget concerns more complex and covert activities than that for intelligence, command, and communications. We have little detail about the central systems and organizations which account for the first three items in the bar representing direct intelligence. Yet we retain them in our proposals undi-

(2) Study of the flood of openly published sources, the books, broadcasts, and papers, the statistics which bear on the forces, strength, and intentions of other powers. Interpretation is of course crucial here.

(3) Signal interception, or electronic intelligence. This is simply the modern counterpart of eavesdropping. It depends mainly on the fact that others, too, have a flood of their own communications in the ether for all who are able to intercept them, mainly from the wide variety of radio bands, or more rarely by wiretap. An important corollary to this big operation—with its many listening points worldwide—is a powerful organization for decoding (and coding too), an extensive cryptologic program.

(4) Photographic images, and their generalization to the use of all sorts of novel sensors for surveillance of the environment. This is the most highly developed of systems which has proliferated in many directions. Land and sea are under constant watch from satellite spies-in-the-sky. They can do everything from monitor crops to notice and report the launch of a rocket. They study and record construction and industrial activity, the movement of ships, trains, even truck traffic; they are capable to some degree even by night. Deep in the ocean waters are large listening networks which report to dozens of shore stations the motion of submarines and ships, especially through the narrow straits from Iceland to the Azores, and beyond the Aleutians, where naval forces must pass to gain the open sea. U.S. instruments listen to the distant sounds of earthquakes and large explosions underground; the U.S. forces maintain a worldwide weather reporting network, quite independent of the civilian system.

---

minished, as the text explains, save for the operations of the CIA, the specialists in covert operations and de-stabilization.

The portion allotted to widespread operational and tactical intelligence, reconnaissance, and signal interception can, we feel, safely be pro-rated to the smaller general-purpose force size we propose.

Budget material from DEFENSE SECRETARY FY 1977, section VA, and FY 1978, section II, ch. IV. Estimates of costs by authors based on space budget of Department of Defense and press reports, plus manpower allocations in MANPOWER FY 1978, ch. VI.

THE PENTAGON'S COMMUNICATIONS NETWORK

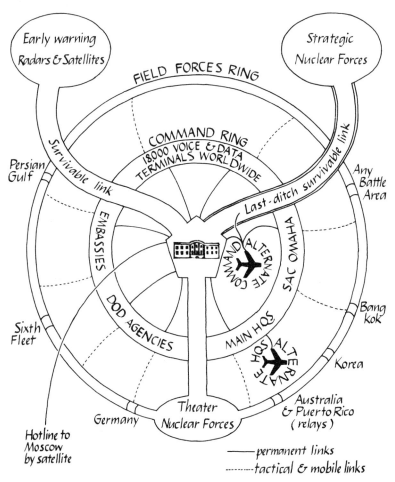

FIG. 60

The diagram, based on similar schematics from the reports of the Secretary of Defense, offers an impression of the architecture of the elaborate data and command network of the Pentagon. Provision for special protection (against jamming and physical damage) of the links to the strategic warning and offensive forces is conspicuous, together with provisions for alternate command centers at various levels.

All links shown are realized in appropriate ways, by land lines, radio, air-

"Communications, command, and control" are less diverse, but they also are complex and remarkable. They link the U.S. forces by voice, teletype, picture, and computer-data channels literally around the world and beyond to space. This system is schematized in a drawing of three rings, the President and Secretary of Defense at their center (Figure 60).

One important consideration, especially important in the nuclear era, is the problem of maintaining some minimum link even after sudden overwhelming nuclear attack. To this end a strange shadow system has come into being, a set of alternate centers, mobile or hidden, airborne and underground command posts, which can take over command should the normal headquarters suffer crippling damage. For example, three big radio stations around the world are designed to transmit information to the nuclear-missile submarines using waves suited to penetrate the shallower layers of the ocean, so the boats can hear messages—slow and halting messages, to be sure—while remaining submerged. Those stations might be destroyed by nuclear attack. Therefore they are supplemented by a set of 15 long-range aircraft specially equipped with powerful transmitters and able to dangle antennae miles long, ready to take over from the ground stations.

Of course, all these channels are subject to jamming and interception. Much ingenuity is spent to inhibit such actions by others, and by the same logic, to prepare operations of our own of the same sort. Security against such efforts and counterattack is a major and shifting charge on the systems, down to the level of tank divisions and tactical aircraft squadrons.

To give some more sense of this intricate and little-known world, we mention that the Department of Defense has a network of several specialized communication satellites in orbit, and rents a good deal of civilian capacity as well. Satellites have many other uses; communications relay, constant surveillance of missile launch and of other satellites in space, watch over space-borne nuclear

---

borne and ground radio relay, and half-a-dozen communication satellites, generally offering redundant possibilities.

The destinations shown on the outer ring are meant as examples of the shifting tactical needs of the field forces.

See especially DEFENSE SECRETARY FY 1977, diagram on p. 223, and discussion in the report, as in that for FY 1978, pp. 247ff.

weapons tests anywhere this side of the sun, weather forecast, and cloud-cover sampling on a rapid basis, immediate pictures by day (and at night by infrared) of ground and sea surfaces everywhere, and increasingly widespread and precise means of giving ground positions to ships, planes, and even to missiles in flight. This activity accounts for almost half of the total U.S. expenditure in space, a quiet rival to the public and civilian National Aeronautics and Space Administration, probably under the National Reconnaissance Office (NRO). The U.S.A. is not alone in orbit; the U.S.S.R. has launched hundreds of surveillance satellites as well, usually rather short-lived, and also maintains some level of constant surveillance of the seven seas by radar from orbit.

The large-scale ground support by the National Security Agency (NSA) or the NRO, and its various military inputs are not fully known publicly as to size and scope. Most of the signals interception itself is done by military personnel; indeed, the degree to which the various specialized systems—say, the Navy Sound Surveillance Under Sea System, taking notice of all motions in and under the sea—is represented in this budget category is not clear. There is no doubt that most of these expenditures are carried out under authorizations closer to the operational commanders. Such forward "intelligence-related" systems are found throughout the forces, for reconnaissance and eavesdropping. They include aircraft, ships, and many land positions equipped to gain knowledge of the opposing forces and activities by direct and indirect means.

It is hard to grasp the sheer bulk and constant flow of all these intercepted messages and the ceaseless images. Tens of millions of documents, thousands of miles of recorder tape, millions of video images are made, often never to be translated, transcribed, or examined. The tactical material is frequently redundant with similar efforts by NATO allies in the same sector, several national services carefully picking up and recording one and the same Soviet transmission, of transient interest. The prodigious ability to collect and to store drives the system to excesses of commission; nothing is left undone, because any scrap *might* prove worthwhile!

How can we approach this complex and half-hidden activity under the principles which frame our entire work? We expect this brief statement of intention and the rough estimates of expense which follow to be an adequate beginning.

It is far from clear that the vast array of space-and-air images of all parts of the world, the flood of intercepted messages, and the tens of thousands of people engaged in examining them with the remarkable systems centered at NSA (and NRO?), are daily needed for our security. There is truth in the argument that the level of detail represented is beyond the capabilities of any political leadership to grasp. A State Department or a National Security Council may come to rely—indeed probably it ought to rely—on a more aggregated and less volatile estimate of the tendencies of other nations than on daily plots of troop and ship disposition. The present surveillance seems closer to a state of actual war than to one of prudent preparation in a world seeking peace by design. On the other hand, the changing facts of the world distribution of forces can be no mean aid to reasonable policy, to weapons planners, or to military deployment.

The growth of our dependence on "national means of verification" of signed agreements is real, not easily to be pruned back. The existence of satellite verification for counting missile sites was codified in the 1972 strategic arms agreements. We therefore take a prudent position: We here propose to maintain during the first years of the new defensive stance the full present capability of reconnaissance, environmental survey, and cryptological and signals-intercept capability, at all levels reporting to the central agencies. We propose only that the widespread networks which locally collect tactical information—recording local radio from Warsaw Treaty army units, for example—which they report to commanders beneath the central level, be allowed to decrease gradually. Year by year, they can fall in proportion to the declining overall level of those forward forces. For instance, the field signals intercept stations in our Asian forces would go toward zero as those forces are withdrawn. The stations might be turned over to local forces, as has been partially done on the East-West German border. Any need for our own survey of the Koreans, for example, can be carried out by airborne or satellite-borne sensors, plus open intelligence and the covert means of espionage already present. Since we will describe only overall budget rubrics, we are confident that interaction and jurisdictional problems can be easily settled within the overall effort we envision.

We present here in two parts a crude breakdown of expenses in this entire domain, and our proposal for a prudent rescaling, after say five years of steady decrease toward a safer and more measured level.

The two parts are these:

    (1) The portion of the effort now budgeted for the program called *intelligence and communications*, plus some esti-

mate of the classified portion. We emphasize the part which reports to higher commands. These figures are directly published.

(2) Our estimate of the portion of such intelligence and communication which takes place at local, tactical level. The personnel and equipment for its work is now budgeted within the programs directly concerned, in the general-purpose forces. For example, this would include the intelligence companies and the electronics warfare companies now found in regular Army divisions. It also includes the reconnaissance aircraft attached to tactical air wings.

How manifold these activities are can be grasped by the organizational tables published. There one sees the National Security Agency (and the C.I.A.) off to one side, and on the other the three services, each with an intelligence agency of its own. The Joint Chiefs have their own D.I.A. (Defense Intelligence Agency), there is a cryptologic branch in each service, and of course local commands repeat the pattern. How much is redundant, how much functional is beyond our power to say. There are at least seven intelligence-related activities, as tabulated in Figure 59, such as direct-support signals-intercept, for example, which are tied "more explicitly to combat force readiness than to a consolidated intelligence function." It is such forces which should drop in effort and cost as the combat forces go down; the central functions we have chosen to retain at strength.

Most of the actual reduction in the intelligence and communications budget would occur in C.I.A. programs and in that direct activity which is in the forward zones, closely related to the deployed strength of the various armed forces. This proposal retains a central intelligence and communications expenditure at a higher proportionate level than the present budget. We see this as valuable reassurance for those who otherwise might fear the unexpected during the first few years of a new defense-centered policy. The margin we allow here is unusually generous, because it arises out of a three-fold barrier of the unknown. We do not know the nature and real utility of the current intelligence product: We do not know much about the arms and intelligence technology yet to come; we do not know well how to judge future developments by our adversaries.

We have therefore chosen to maintain this very large, redundant, and overactive capability as a *hedge against uncertainties.* Yet all the burden of that uncertainty ought not to be borne by our side, the side of decrease of tension; no one can argue that the present U.S. active position is in any deep sense secure; quite the contrary. We therefore invite novel proposals for other dispositions of these remarkable facilities and organizations. It does not seem unreasonable to think of placing a share of them at the service of international peacekeeping, or even of international economic efforts, from marine communications to resource studies and disaster aid. Our national goal is not military strength in itself, for that is at best a means. Our true goal is, after all, national security in peace.

14

GENETIC APPARATUS OF THE ARMS RACE

The future direction of the arms race is determined in the large part by current expenditures on research and development (R & D). Although this category accounts for only about 10 percent of the total Department of Defense budget, it is worth examining closely, for from it come the temptations to enter into spirals of unsafe technological escalation. A careful policy of control of R & D can do much to inhibit the arms race while still providing protection against possible technological improvements by our adversaries.

An examination of worldwide military innovations makes it clear that the United States R & D has forced the pace. Most significant research and development advances in the past 30 years have been made by the United States, and in terms of both productivity and ingenuity, the American military R & D community far outstrips the U.S.S.R. Within the U.S.A., military R & D is much larger than the Federal support of essential research on energy and health (Figure 61). The need, therefore, exists for the United States to de-escalate this critical component of the arms race by slowing down the rate of advance in military technology.

The stages of the research and development process are described in Figure 62. We will discuss the early stages separately from the later stages since the processes and implications are somewhat different.

EARLY STAGES: RESEARCH AND EXPLORATORY DEVELOPMENT

The first two stages are called "Research" and "Exploratory Development." They provide the technological basis for new weapons and new military concepts. The number of projects

FEDERAL SPENDING ON RESEARCH & DEVELOPMENT

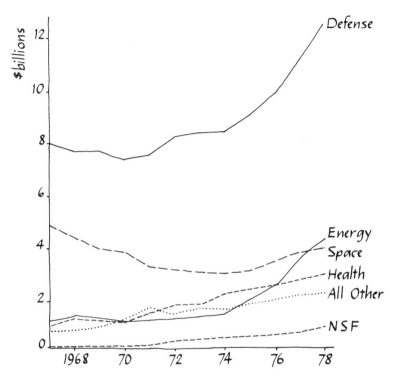

FIG. 61

More federal dollars are spent on defense research than on any other area of R & D, and these expenditures are growing faster than the rate of inflation. Even such national priorities as energy and health research lag far behind and show no sign of catching up; the National Science Foundation shows the lowest level of funding.

The fiscal 1978 budget of President Gerald Ford as illustrated in *Aviation Week and Space Technology,* v. 106, n. 6 (Feb. 7, 1977).

undertaken is very large (20,000 distinct topics are provided for in the current budget) and cover an astounding diversity of activities. Examples include basic nuclear and laser physics, the design of new transistors to replace expensive and short-lived transmitter tubes,

FUTURE COSTS OF A NEW WEAPON CONCEPT

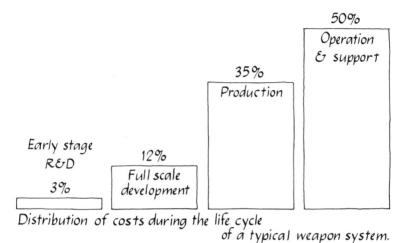

Distribution of costs during the life cycle
of a typical weapon system.

FIG. 62

The early stage R & D which eventually gives rise to a new weapon costs little compared to the subsequent expenses of developing, producing, and operating the weapon. Every million dollars spent on basic R & D for military purposes means a commitment of many more millions once the successful products of that early stage are developed and realized. Attractive concepts, even rough designs, are cheap to produce, but they provide temptations to high budgets and new risks years later.

The distribution of costs through a life cycle of a typical weapon system is derived from Figure 1 facing page III-3 of PROGRAM RESEARCH FY 1976. The reader should consult that report for exceptionally lucid detail on the R & D program.

investigation into theoretical military concepts, and more effective software or production schemes, the development of cheaper titanium sponge, research into tropical diseases, and the surgical care of injured subjects, and even a remarkable instance in which commercial plumbing supplies were adapted to make hand-held missile launchers. The variety is all but incredible, ranging from the subtle results of contemporary science to commonsense ideas on making handier gadgets of every sort.

Most research and development work in these early stages is not committed to any particular weapon system. For example, theoretical studies of supersonic wing design might be made in these early stages, which although they might eventually be used in a program such as the B-1 bomber, would not be designated for that purpose at the time of the research. Although many items are funded, the costs per item are small because ideas are cheap and large-scale models or mock-ups are not constructed here. Only a relatively small number of people need to be supported to generate the ideas and new concepts which are the products of these early stages.

Out of the total research and development budget of $11.9 billion in FY78, about $1.9 billion (or 16 percent) is budgeted for the early stages of R & D. This 16 percent, however, has far-reaching effects on every military mission, and a single item, such as an improved transistor, may effect all weapons systems if one looks far enough ahead. These early stages are also heavily supplemented by civilian research and development. Governmental and foundation support of basic science, and industrial expenditures on research and development, all contribute to the technological base which supports the military effort. Laser physics is one of the many examples of an area of military applicability that was originated in the civilian sector. Cross-fertilization can also occur in the reverse direction and some military supported research has spinoffs which benefit civilian research and development. Examples of this include the jet engines now used in commercial airliners and the early developmental work on computers in the 1950s.

FROM RESEARCH LABORATORY
TO OPERATING SYSTEM

Once an innovation has been found feasible or promising there are still many steps that must be followed before it can become part of an operating system. Thus, while improvements of supersonic wing design might be a product of early stage R & D, the utilization of the new concepts in the design of a supersonic aircraft such as the B-1 bomber would not usually come until the later stage research and development work. These stages are usually referred to by the Department of Defense as "Advanced Development" and "Engineering Development" (see Figure 63). In these stages prototypes for future aircraft are designed, constructed, and tested. Also here occurs the long process of working out the surrounding details which makes the system function in the real world. In these stages

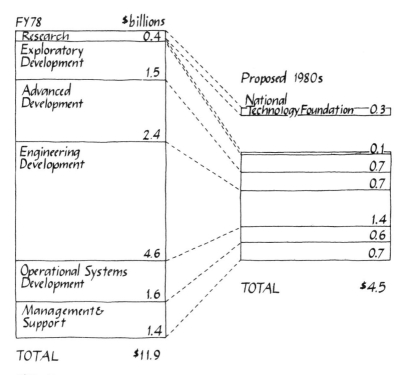

MILITARY R&D BUDGET PROPOSALS

FY78	$billions
Research	0.4
Exploratory Development	1.5
Advanced Development	2.4
Engineering Development	4.6
Operational Systems Development	1.6
Management & Support	1.4
TOTAL	$11.9

Proposed 1980s

National Technology Foundation — 0.3

0.1
0.7
0.7
1.4
0.6
0.7

TOTAL $4.5

FIG. 63

The dollars (in billions) the Defense Department spent on each stage of research and development in 1978, and our proposed budget for the 1980s, are shown above. Seventy-five percent of the first stage of basic research is transferred from military to civilian control in our proposal. With these funds we would form a National Technology Foundation so that the investment of the U.S. in basic technological research will be more flexibly responsive to national needs. Military-related R & D would be restrained especially in the later stages, so that the genesis of new weapons is slowed to a safer rate.

DEFENSE SECRETARY FY 1978; and PROGRAM RESEARCH FY 1978.

there is no quantity production or training; funding and development only proceeds up to the preproduction models. Once the new weapons system is approved for production, as was the B-1 bomber to a limited extent in 1976, the costs for the research and development support of further modifications are found under "Operational Systems Development," where they are accounted for under the budgets of the military forces.

These R & D funds are divided among universities, federal contract research centers, Department of Defense laboratories, and industry. Only about 3 percent goes to support university research and the majority, even in the early stages, is conducted in industry.

Another way of looking at these expenditures is to examine the costs of development of the typical weapons system over its total lifetime. This process is illustrated in Figure 62. It can be seen that there is a considerable amplification effect on future budgets from work done in the original research stage. An idea which seems attractive and does not cost much to generate can be extremely expensive to implement. It is this effect which constitutes a major determinant of military budgets in future years.

POLICY RECOMMENDATIONS: A MINIMAL BUDGET FOR RESEARCH AND DEVELOPMENT

We have considered a number of ways to cut back in the R & D effort while minimizing the security risk to the United States. One policy would be to maintain the support of the early stages of research and development at their current level, but drastically cut back the later developmental stages. This would allow the United States to maintain a lead in basic know-how and concepts, but at the same time control their application. Although superficially appealing, we doubt the effectiveness of this approach. New ideas from the laboratory, especially those that appear to save money or greatly improve some weapons system, may be difficult to resist. More importantly, it is difficult to keep the wraps on a new discovery, especially if it has general applications. As soon as some other nation picks up on the idea and starts to develop it, the United States will be forced to follow suit.

Limits must therefore be set on early stage research and development as well as the later stages. In considering early stage research and development it is clear that there is a subtle gradation from projects which may be equally applicable to civilian or military purposes through to those which are tailored specifically to military problems. With many early stage projects it is impossible to predict with certainty the ultimate application. Moreover, the diversity and large number of individual projects make generalizations hazardous. Thus, we recognize that in order to implement the policy we recommend, a detailed study of every project must be undertaken bearing in mind the problems of prediction we have discussed.

In setting policy it must be realized that spending on research has a relatively greater effect on future budgets than on military superiority. Technology is only one factor in a wartime situation. Leadership, morale, strategies, intelligence, numbers of trained men, and amount of equipment also play important roles. One lesson of Vietnam was that overwhelming technological superiority is no guarantee of military success even against an enemy with minimal technology and no R & D budget.

Constant spending on R & D and the replacement of existing weapons systems as soon as innovations become applicable is an expensive way to approach technological improvement. Because of the need to change procedures and retrain staff following the introduction of each new system, it may be often just as effective, and certainly cheaper, to wait a generation in the development of new technologies before applying them to operational use. This consideration applies particularly to technologies which are rapidly developing, such as computers.

We also realize that too great a cut in research and development may be just as undesirable as no cut at all. Some hedge is required against technological development by military rivals which may upset the curent military balance. Fortunately, the lead time from basic innovation to application is considerable. It may be 10 years or so for the developer and somewhat less for the second nation to enter the field. Major scientific innovations are usually universally recognized throughout the world's scientific community and so it is the application of the innovation to a viable new weapons system that must be detected. Provided military intelligence on technological developments is reasonable, the timespan should be sufficient for the greater U.S. management, technical, and financial resources to overtake any adversary. However, from time to time, even now, there are liable to be occasional gaps in specific fields. Provided that there is no disparity overall these should neither be surprising nor frightening. Only a significant new source of danger should be a sufficient signal for greater gearing up of a research and development effort, not some revision of an enemy weapons system which has little impact on the United States *defensive* (in the strict use of the term) capability.

Let us turn now to our specific policy recommendations, bearing these caveats in mind. Because of the different goals and impacts of early and late stage R & D, we will consider our recommended policy for these stages separately.

Early Stages of Research and Development.

1. We recommend that the support of basic science be removed from the Department of Defense budget. While we recognize the major role the military research and development funds have played in the development of many "high technology" innovations, we believe that fundamental research and development is much better handled by civilian agencies which do not restrict research support to projects with a high probability of military applicability. The basic research process as much as possible should be open to critical review of peer groups. Under these circumstances the product is more likely to be of high standard and also be less expensive than that conducted under heavy security restrictions and supervised by a single contract officer. Only when the nature of a research project clearly requires security clearance or becomes specifically oriented toward a readily identifiable military problem should it be funded and supervised by the Department of Defense.

To implement civilian support of research and development we propose the formation of a "National Technology Foundation." This would stimulate research and development in areas of expensive high technology where private industry is not yet willing to undertake the work. Out of the $400 million now allocated for basic research we propose that 75 percent or approximately $300 million per year be allocated to the National Technology Foundation. We suggest the remainder should be left in the Department of Defense budget to provide a skeleton of basic research support for engineering laboratories concerned with later stage system development.

2. We propose that research and development efforts which are clearly contrary to a stated policy of noninvolvement in the third world (e.g., aircraft carrier technology) or are concerned with the development of weaponry internationally regarded as particularly indiscriminate or inhumane (such as chemical and biological warfare, incendiaries, and maiming weapons) should immediately be halted.

3. We recommend a major cut in the exploratory development budget (the second of the two early stage development levels). We

propose to cut the $1.5 billion in this category by 50 percent because of the need to reduce the rate of military technological innovation to a level that minimizes the amplification effect on future budgets.

Late development stages.

The purposes and implications of late stage R & D programs (advanced development, engineering development, and operational system development) are easier to identify than the early stage because they are largely concerned with development of specific weapons systems. Once the military aims of these new weapons are known, it is possible to review the necessity and desirability of each in the context of our overall policy recommendations.

We recommend a major cut in the late stage R & D projects. As we evaluated each individual project we followed a number of guiding principles in deciding whether to cut the project, partially fund it, or fully fund it at the current level. In some cases we have increased spending because of the different emphasis requirements of our new forces.

1. Certain types of weapons systems we have shown to be incompatible with our policy recommendations for force levels. R & D work which is devoted to such weapons systems has been cut since it no longer will serve any purpose. Examples of this include our recommended termination of the Minuteman II and Minuteman III programs, the MX Advanced ICBM, the Trident program, and the B-1 Bomber.

2. R & D work which is liable to induce a research and development response by the enemy which will negate any temporary advantage should not be pursued. Of particular current concern are the antiradar devices and cruise missiles.

3. Great care should be taken in the duplication of military spending. While there may be some advantage in early stages of the development to have competing similar projects, it is clear that this should not apply to appropriations of major weapons systems. Often there appears to be little justification for certain weapons systems other than the wish to maintain equality, or more accurately repetition, among Army, Air Force, and Navy.

4. We support an increase in research and development spending on relatively cheap weapons which improve defensive capabilities by having an "equalizer" effect. Examples of these would include the antitank and antiaircraft weapons.

5. R & D efforts in intelligence activities need to be maintained to provide a safeguard as we reduce force levels. Thus, for example, we have maintained funding for reconnaissance satellite programs.

Maintenance of research and development capability.

Totally shutting down major sections of high-technology defense industries could present significant problems. Important collections of theoretical and technical know-how are liable to be dispersed when the research and development team is broken up. We believe the best solution is to maintain small highly trained cadres working on real or theoretical projects. From such a base it would be much easier to reestablish a fully operational team should the United States need to rebuild any sector of its military research and development capability. This consideration is part of the justification for our leaving in our budget some projects with low impact on the arms race.

Summary of research and development budget proposals.

Because of the very large numbers and diversity of the projects which are funded by the research and development budget, the easiest way to aggregate our policy proposals is in the form of a budget. In Figure 63, we present our proposals broken down by development stage. We have analyzed in some detail the individual R & D projects funded under each development stage to produce this recommended budget. For a further discussion of methodology and for other breakdowns of this budget the reader is referred to Chapter 19 where our overall budget is discussed.

15

RESERVE FORCES AND INDUSTRIAL MOBILIZATION

The weapons of the American armed forces are served by some 2.1 million men (and women) now on full-time active duty. The arsenals and the production plants offer a flood of weapons to augment or replace those now deployed, should crisis or actual war require it. But what of the trained people to serve those newly ready arms?

For a long time the United States has maintained an organized military reserve force paid for part-time duty only, which can hope to supply a needed increase on short order. These amount to some 800,000 men and women (for years 1977 to 1979). Behind these is a much larger pool of persons—several millions—whose names are on file simply as having received military training in the past. These have some legal obligation to serve when mobilized by Congress, an obligation which arises from the conditions of their discharge or early retirement from the military.

Closest to active military service are the National Guard and the elected Reserves. These are the only true reserves in being in the generally understood sense; persons who are currently paid and trained part-time and organized into miliary units with some reasonable expectation of being called upon. Of next importance is a pool of persons whose discharge from active service is fairly recent, under conditions which left them open to recall but who did not undertake any regular part-time service. They are called "Ready Reserves"; they are not paid or trained, and they are not generally organized into military units. (Some of these persons render service with pay for a couple of weeks' annual duty.) The others of the large pool, who also are not now paid or trained, have seen service less recently; their obligations are so hedged about with legal safeguards that they are unlikely to be drawn upon except for a profound and widely felt crisis. Their mobilization would be

THE SIX RESERVE COMPONENTS

	Manpower	Cost in $billions Personnel	Maintenance & Procurement
Army National Guard	390,000	2.3	1.7
Army Reserve	219,000		
Naval Reserve	52,000	.4	.4
Marine Corps Reserve	33,000	.1	.14
Air National Guard	93,000	1.6	.45
Air Force Reserve	52,000		
TOTALS	840,000	$4.4	$2.7

FIG. 64

The six distinct portions of the part-time forces, the Reserve Components, are listed with their present personnel strength and approximate costs. Note that the Army Reserves are by far the largest and most costly.

See especially DEFENSE SECRETARY FY 1978. See also recent CBO reports on reserve forces.

economically disruptive, and politically an act of considerable moment. For example, mobilization of certain Reserve units was used as a signal of vital U.S. concern in Berlin crises in the past. On the other hand, except for a few token or very specialized units, the large Reserve forces already in being were not called upon during the long Vietnam conflict. The impact of taking men with families and with a degree of political influence outweighed the military advantages of finding ready units. A tide of eighteen-year-old draftees flowed to Southeast Asia, while the Reserves stayed home and trained. (One must admit that a newly trained and fit young man is not a poorer bet as an infantryman than is a weekend warrior six or more years older, who has trained for years, but has no combat experience.)

The organized reserves include half-a-dozen distinct elements, each associated with one of the active services. These differ widely in legal status, command chain, and in relationship to government at federal and state levels. The National Guard, for example, though under state call, also forms a major part of the national organized reserve. We comment on these six organizations (Figure 64):

Army National Guard and Army Selected Reserve: The National Guard is closely related to many individual states, whose militia it largely comprises. It has certain legal rights and duties for the maintenance of local civil order, under various degrees of independence from the Department of Defense. The command centers of the reserve Army divisions are within the National Guard, with strong regional ties and traditions (not to mention political clout). The Army looks to the Guard—plus the straight Army Selected Reserve—for eight combat divisions, but with the equivalent manpower to some 16 more divisions. The extra men are organized as increased close support troops and replacements, to begin duty after the division has been two months in the field. There are many specialized units—medical, light armor, engineers, signals, artillery, training, logistics—which are not organized into divisions.

Naval Reserve: There are about 28 destroyer-sized vessels assigned to the Naval Reserve and 30 smaller craft. These often have crews half of which are full-time active naval personnel and half part-time reservists. The ship is maintained in ready condition and can be fully manned part of the time, usually for seasonal training voyages. Naval aviation includes patrol squadrons, plus three wings each of carrier attack and antisubmarine warfare groups. In the Naval Reserve especially there are also Individual Ready Reservists, persons not permanently attached to any unit, who bring their specialized skills to a wide variety of jobs in the fleet and in land support, filling in during their training periods in whatever unit they are sent to serve.

Marine Reserve: These are organized as one amphibious force, a counterpart to the three on active duty; they form one division of Marines plus one air wing, without extensive support.

Air Force: The Air National Guard mans about 40 tactical fighter and interceptor squadrons. They comprise under a third of the active tactical air force, but the bulk of all units currently assigned to strategic interceptor air defense of the continental United States are Air National Guard units, including half of the Delta Darts, our top-line if elderly interceptors.

The Air Force Reserve and Guard maintain 36 tactical airlift squadrons, a reserve force about one and one-half times the active tactical airlift forces.

It is clear that the Air Force and Navy reserve units of all kinds are nearer ready for their more specialized service than are the larger Army components, with their orientation toward ground combat support. Present policy is to try to reduce this distinction; the problem is evidently no easy matter. Moreover, there is no point in large reserve units whose training or manning is not adequate to fit them for quick response in weeks. Since the same training can be given to people drawn from civil life in a few months' time, such units seem of little use. Keeping a small organized well-trained framework for the necessary bigger units might be the better bet.

WHAT THE RESERVES PROVIDE

The chief missions of the Reserves seem to be three:
1. The swift augmentation of active forces engaged in conventional war in Europe. The Army divisions and their backup and the major responsibility for tactical airlift of the ·Air Force reserves are plainly meant to meet this need. So perhaps do the surface escorts and antisubmarine wings of the Naval Reserve.
2. Maintenance of organized structures which could be completed efficiently in time of need. Reserve units could be fleshed out by drawing upon draftees, volunteers, or Standby Reserves, persons with obligations for service, with military training a few years back.
3. Preparation of a body of persons available for ready local use in cases of domestic need, whether for extensive civil disorder or for serious physical emergencies, earthquake, fire, or flood.

We believe that a prudent attitude to the reserves would be to maintain those elements with purely defensive roles—for example, continental air defense. Quick-response capability also seems essential. Already Air Force Reserve units are generally available on short notice. Civilian maintenance crews make it possible for units to deploy within about three days. Present Department of Defense emphasis is to spread this style of swift response to Army units, along with modern equipment. (We still see the World War II structure very strong in the Army Reserves, which include many units with specialized duties such as military government.) It is not realistic to expect that during peacetime U.S. combat infantry units can be kept adequately trained for combat by weekends and brief summer maneuvers. Note that the Army Reserves include only about one-third combat forces, but about two-thirds of the more gradually involved support units, divisional "slices" for replacement of loss and for reinforcement. This reflects the sense we share of the Army Reserve problem. The airlift crews can in fact fly their big planes long distances on weekends, everything real save enemy fire. But it is harder to train combat infantry on a part-time basis.

We propose the following new arrangements for the reserves:

1. Prorate the strength of Selected Reserve—the paid and trained units and individuals—to the new overall active personnel level we recommended. All units which cannot be made ready swiftly, or whose mission is useful only after months of war, should be eliminated. This requires a cut of more than half in Army Reserves.

The smaller active forces we recommend will result in release of a good-sized pool of trained people, available for all the reserve functions and indeed for recall.

2. Hedge the function of slow-response units by the use of cadres; that is, a skeletal framework of key personnel and equipment devoted to training and to preparation for rapid emergency expansion. The level of this cadre force will be about one-fifth the present size, but with somewhat increased expenditure per man; more paid time and newer equipment.

3. Retain Army National Guard divisions close to the cadre level, but allowing each state to set up the forces it deems necessary. For a long time no state callup has exceeded the 13,000 men deployed in Los Angeles in 1965. Thus, it would seem plausible to cut the combat-intended divisional Army National Guard down to one-third, somewhat stronger than the cadre fraction; support elements seem suited for the cadre treatment, a cutback to one-fifth.

4. Keep reserves as a whole away from nuclear weapons. These surely demand fully trained crews.

5. Consider the use of reserves for long-time maintenance of some of the new equipment which we recommend for removal from the active forces.

6. Use reserve components to maintain a readiness force for a few special operations heavily cut by our proposals: for example, carrier-attack forces, and amphibious-landing task forces. This would make it possible to reverse the cutback without long periods of training, should reactivation ever become necessary. Carrier operations, for example, require pilots, support personnel, and aircraft. If these programs were totally stopped, it might take years to reestablish the skills and structures. One-tenth of the sums removed from active service operations could maintain these

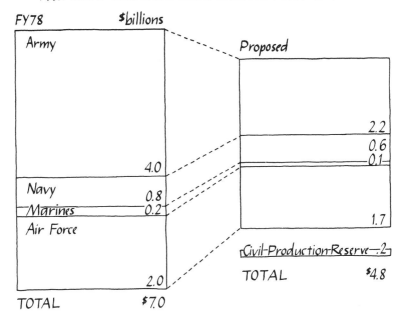

PROPOSED RESERVE COMPONENTS FOR THE 1980's

FY78 $billions

Army

Proposed

2.2
0.6
0.1

4.0

Navy   0.8
Marines   0.2
Air Force

1.7

2.0

Civil-Production-Reserve—.2

TOTAL $4.8

TOTAL $7.0

FIG. 65

The present and proposed budgets for the part-time forces are presented, divided into the services. The new proposal, for a Civil Production Reserve, is a suggested economical method of retaining the ability to reverse sharp reductions in certain weapons systems, by keeping key industrial production possibilities in reserve (see text).

DEFENSE SECRETARY FY 1978.

abilities qualitatively, but on a small scale, over a five-or ten-year transition period. (This particular example is not in fact essential if a few active carriers are retained.)

    7. We suggest tentatively an entirely new form of reserve operation, suited to the overall posture of our present recommendations. It is a Civil Production Reserve.

    Here is its function. Many complex or specialized military products are not found in civil life. If their procurement is cut off entirely or below economic scale, the production stops, and designers and craftsmen drift to other jobs. It might take a considerable time to reassemble such an industrial skill again, were it needed for a changed military situation. Under our proposal, cadres would be maintained at chosen key development shops and factories, people paid and equipped to work on design improvements and prototypes but *not* on major new technical improvements (see Chapter 14 for a discussion of R & D). Firms which made carrier landing equipment, for example, could be kept operating at a modest level. Their staff and shops would be reduced or diverted for the most part to other tasks, but about one-tenth could maintain their military engineering and production skills.

The substantial force reduction for the Reserves overall would maintain their functions at a level proportionate to the armed forces we recommend in general. At the same time we provide for the preservation, over a considerable transition period, of certain activities which might otherwise be claimed to limit the possible rate of rearmament, should events somehow require such reversal. We believe it is relatively easy to provide safely against contingencies of that sort by modest and inexpensive innovations, entirely consistent with a substantial overall reduction of forces. The principle goes beyond the problem of the reserves and may well turn out to have other applications. In Figure 65 we summarize rough costs of this overall position.

16

SUPPLY, TRAINING, AND ADMINISTRATION

One-quarter of the defense budget is spent on a central supporting framework which underlies the various combat forces and weapons systems described elsewhere in this book. Such ongoing operations as central planning and administration, weapons and equipment maintenance, the basic training and medical care of military personnel, and the transfer of men and supplies to and from bases located around the world account for about $30 billion of the FY 1978 defense budget (see Figure 66). (A further $9 billion is spent on another general personnel item, retirement pay, but this is treated separately in Chapter 18 because of the special consequences of our proposals for the rate of retirement.)

What first stands out about the central support functions is that they involve very little investment in major equipment or facilities—only $2 billion out of the $30 billion. Most of the funds are spent on the pay of military and civilian personnel used in administrative, maintenance and training tasks, and the fuel and other consumable supplies for day-to-day operations. This is in contrast to the combat force oriented sections of the defense budget, where over 50 percent of the funds are spent on major weapons, equipment and facilities, and correspondingly less on personnel and operational costs.

Second, large numbers of civilians are employed in these support activities (Figure 67). About half of the one million civilians employed by the Department of Defense are employed in administration, maintenance, and training. They store and manage supplies, feed, clean and maintain, repair, service and operate, in each case replacing military personnel who are scarcer resources. They work both domestically and overseas. The other half of the civilians are assigned directly to the combat forces—but again serve in similar functions. Certain specialized support functions such as those of the independent Defense Agencies, which are discussed separately

SOME BASIC SUPPORT COSTS

	Maintenance	Training	Administration
Army	3.4	5.5	0.6
Navy	3.8	3.3	0.5
Air Force	3.5	3.2	0.4
Marines	0.2	0.7	0.1
Defense Agencies	1.0	1.0	0.3

TOTALS in $billions	$11.9	$13.7	$1.9

FIG. 66

Almost $30 billion—one quarter of the defense budget—is spent on mainte-
nance, training, and administration. The Army, with more personnel, spends
more on training and administration, but the costs of maintenance, both per head
and in total, are greater in the Navy and the Air Force, with their more complex
and expensive equipment.

DEFENSE SECRETARY FY 1978, as amended by the 1978 Carter budget.

below, are performed almost entirely by civilians. The annual costs
of civilian employees (pay, insurance, and retirement benefits)—
paid partly from support programs and partly from combat force
programs—amounts to about $19 billion, or $18,500 per person on
the average. These funds are provided in the appropriation for
operations and maintenance.

The expenditures on the support functions vary among three
services (Figure 66). The Army, with its large numbers of personnel
(about 800,000) as compared to the Navy and Air Force (about
550,000 each) spends more than the other services on training and
medical care. Of the funds allocated to personnel support nearly
half are in the form of pay of military personnel, either the pay of
those giving training and medical care, or the recruits.

In contrast, the costs of maintaining and transporting major
weapons and equipment are about evenly divided among the three
services. The Army has more individual items of equipment but
they are on the average smaller than the fewer but enormously
complex weapons systems of the Air Force and Navy (consider the

A MILLION CIVILIANS AT WORK

	Army	Navy & Marine	Air Force	Defense Agencies
Supply, Logistics & Base Operations	261,000	80,000	188,000	51,000
Medical	25,000	10,000	8,000	
Research & Development	21,000	35,000	19,000	2,000
Naval Shipyards, etc.		148,000		
Training	19,000	14,000	8,000	
Miscellaneous	54,000	34,000	34,000	26,000
TOTALS	380,000	321,000	257,000	79,000

FIG. 67

The table gives some sense of what the million civilians employed by the Department of Defense actually do. The Naval Shipyards, establishments with a long history, are listed separately; there are no comparable Air Force factories building aircraft, for example. Repairs, maintenance, and conversion are carried out for all services, both by private industry under contract and by employees of the DOD. The people so employed are included in other entries, as far as they are employees of the Department of Defense.

MANPOWER FY 1978.

gargantuan aircraft carriers or highly complex swing-wing aircraft). Most of this maintenance involves centrally performed overhaul of equipment in special plants, which, unlike the day-to-day maintenance at the combat unit level, is largely carried out by civilian rather than military personnel.

DEFENSE AGENCIES

Another type of support, funded separately from the three services, is provided by the 15 independent Defense Agencies. These agencies range in size from the large to the very small. The Defense Supply Agency is concerned with the logistics of supplying materials, food, clothing, and ammunition to the military forces around the globe. Overall military planning and policy is the province of the Joint Chiefs of Staff, an office funded as a small separate agency. The National Security Agency operates in the world of electronic communication, coding and decoding messages transmitted among bases, stations, ships, and aircraft. (Its size is secret; perhaps 20,000 persons?) The Defense Nuclear Agency works in conjunction with DOE (Department of Energy) in testing and overseeing the production of nuclear weapons and monitoring foreign nuclear tests. Specialized map-making services are provided by the Defense Mapping Agency. Basic R & D, or that on complex new systems and concepts in an early stage of development and therefore not yet assigned to an individual military service, are conducted by the Defense Advanced Research Projects Agency. Most of the DARPA projects concern sensitive items developed in a research effort which provides a direct support for the Office of the Secretary of Defense.

RECOMMENDATIONS

In keeping with our overall force reductions we propose a reduction in the size of the basic support framework. However, rather than a 40 percent to 50 percent cut proportional to the rest of the forces, we would propose a lesser reduction of approximately one-third. Our reasons for this are several.

First, we do not wish to weaken in any way the capabilities or state of training of the trimmed-down forces we have proposed in this book. Relatively more attention if anything should be given to logistical support and medical care. The morale of these forces should be high and the recruitment and support of these forces optimal.

Second, within the complex support framework there may be economies of scale. It is unreasonable to expect a fully proportionate cut in the framework, except in areas where the organization is inefficient. Certainly the reduction in the size of the armed forces and in the deployment of new weapons will require a sizable cut in the supporting structure, but certain agencies must still provide similar services to the reduced military forces.

Third, the nucleus of a large wartime force support framework (we hope never needed) is required. An example is the medical units which are already disproportionately large for the current force

size. The peacetime slack is taken up by the provision of medical services to civilian employees and dependents who would not be accommodated in wartime. This seems a sensible scheme, and might apply to basic training, supply, and maintenance cadres also. Finally, we do not know precisely the amount of support and the level of detailed organization needed for different size combat forces. We believe this area requires further study and recognize that part of the move toward a more limited active military force should involve continuing studies of the needed support elements. It is possible that a cut in support costs, proportionate to the cut in combat forces, would be in order—or, if there are currently inefficiencies in the support framework, that a greater than proportionate cut might be made. Until better information is available, however, we prefer to err on the side of caution, cutting less than proportionately, rather than risk erosion of the readiness of those forces we retain. In any event, large-scale management, rather than military or foreign, policies are here in question.

17
MILITARY AID
AND TRADE

The international trade in nonnu'clear weaponry is considerable.
Approximately 150 countries—about all of them—import weapons
while as many as 25 produce and export military hardware on some
scale. The international arms trade now amounts to over $20 billion
annually, a drastic increase from the $300 million in trade 25 years
ago: this is a steady increase from as recently as 1970 when world
arms trade was under $10 billion. There has been a dramatic rise in
arms exports by the United States over this period—from $2 billion
to $10 billion. The variety of weaponry flowing is broad—from
simple handguns and rifles to highly sophisticated jet fighters, naval
vessels, missiles, and remarkable electronic gear.

In the fiscal 1977 and 1978 annual statements of the Department
of Defense, the small amount of space devoted to the arms trade, 10
pages out of 271 and nine pages out of 350, respectively, belies the
importance of worldwide weapons transfers. The manner in which it
is presented in the defense budget also understates the volume of
trade. In defense budget jargon, arms transfers are referred to as
either "Support of Other Nations" or "Military Assistance Programs." The totals for the last two years for these two categories,
which are not mutually exclusive, are $1.3 billion and $1.1 billion,
respectively, for fiscal year 1977 and $1.3 billion and $1 billion for
fiscal year 1978. These figures understate current U.S. worldwide
arms exports by a factor of 10 due to exclusion from defense budget
figures of all weapons exports not shipped on credit or subsidized
terms. Ignored are the transfers on a direct purchase and cash basis,
now the large majority.

Our analysis is concerned with the total volume of U.S. arms
trade. While this approach includes much more than do past
defense budget tables, we believe it allows for a fairer, more helpful
analysis. The most recent 1978 defense budget discussion ac-

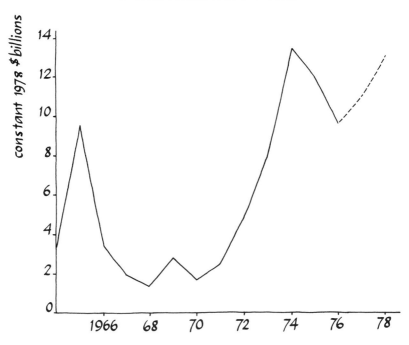

FOREIGN MILITARY SALES

constant 1978 $billions

FIG. 68

American foreign military sales show a marked increase since 1970, mostly due to sales to the Middle East. This has made the United States the Number One arms salesman in the world. There are also commercial and grant transactions, not included here, which would raise these figures still higher. It is estimated that foreign military sales rose in 1977 to over $11 Billion, and in 1978 to over $13 Billion.

DEFENSE SECRETARY FY 1978.

knowledges this by including some account of total U.S. arms sales abroad.

International arms transfers involve a variety of transactions. Suppliers and recipients include both official governmental agencies and private groups as well. Weapons are transferred under many types of arrangements: cash sales, credit, surplus sales, outright grants, and barter agreements. The United States, for example, has

US WEAPONS CUSTOMERS

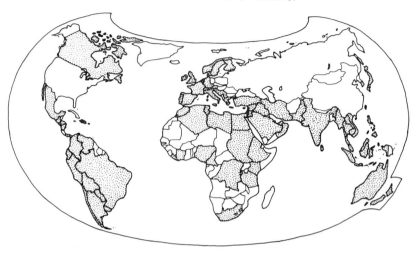

Canada	Austria	Algeria	Egypt	Burma
Argentina	Belgium	Cameroun	Iran	Cambodia
Bolivia	Denmark	Ethiopia	Iraq	Indonesia
Brazil	Finland	Ivory Coast	Israel	Japan
Chile	France	Kenya	Jordan	Laos
Colombia	Greece	Liberia	Kuwait	Malaysia
DominicanR	Ireland	Libya	Lebanon	Philippines
Ecuador	Italy	Morocco	Oman	Singapore
El Salvador	Luxembourg	Nigeria	Saudi Arabia	S Korea
Guatemala	Netherlands	S Africa	Syria	S Vietnam
Haiti	Norway	Sudan	Yemen: Saña	Taiwan
Honduras	Portugal	Tanzania		Thailand
Jamaica	Spain	Tunisia		
Mexico	Sweden	Zaire		
Nicaragua	Switzerland	Zambia	India	
Panama	Turkey		Pakistan	
Paraguay	UK		Sri Lanka	Australia
Peru	W Germany			New Zealand
Uruguay	Yugoslavia			
Venezuela				

FIG. 69

American weaponry has been sold or given away to over 50 percent of the world's countries during the past decade. The recipients include most developed countries except for the Soviet Union and East European nations, and the majority of the developing countries in Asia, Latin America, and Africa. This transferred weapons to foreign recipients through diverse official programs such as Foreign Military Sales, Military Assistance Program, Security Assistance Program, and Military Assistance Service Funded. Some of these programs are included in the U.S. defense budget, while others have their own appropriation. Some military aid programs even fall under the rubric of "economic assistance"; one such program was the "Food for Peace" effort which has been used, misleadingly although only indirectly, to finance the arms transfers and military development of certain less developed countries. It is very difficult, however, to estimate· the amount of economic assistance used as military aid; such figures will therefore not be included in this study except to point out that they exist. On the whole, grant military assistance was larger than sales until 1974 when sales slowly replaced grants.

The complexity of international weapons transfer arrangements, as well as the difficulty of accurately pricing arms on the international market, does not lend itself to simple statistical compilations, comparison, and bookkeeping. For these reasons, there are widely differing estimates of U.S. and world arms transfers. It is nevertheless possible to make some broad and important observations on the volume of U.S. arms transfers since 1950.

Deliveries of U.S. military hardware abroad rose quite sharply after the Korean War and then fell gradually until 1960 when they leveled off. As U.S. involvement in Vietnam expanded through the 1960s, so did U.S. arms transfers consisting in large degree of grant shipments to Indochina. As the Indochina involvement wound down, however, U.S. arms transfers jumped remarkably, primarily a result of U.S. involvement in the Middle East. From under $2 billion of sales in 1970 and under $5 billion in actual deliveries, the U.S. arms trade had risen to over $10 billion annually by 1974, as shown in Figure 68, and has stayed there since.

One should note that this figure includes only one government category—foreign military sales orders filled directly by our government; the totals would be still higher if military grant aid and commercial military orders, as well as some fraction of economic assistance, were included here. For example, in fiscal year 1974,

---

points to the importance of U.S. military aid and trade in American foreign relations worldwide. It also points to the large amount spent annually on trade in weaponry throughout the world—over $20 billion.

ACDA: *World Military Expenditures and Arms Transfers 1966–1975.*

the figure shows $10.6 billion in foreign military sales orders; this
total would be raised to $12.4 billion if one were to include the $0.8
billion in military assistance and $1 billion in commercial military
orders for that year.

The U.S. total in arms sales since 1950 is over $100 billion in
then current dollars; in 1976 dollars this figure would be still much
higher, averaging about $6.5 billion annually for a total of over
$150 billion. There have been over 80 customers worldwide, as
shown in Figure 69, for U.S. weapons. In fact, very little of the
world outside of the Soviet Union, China, and Eastern Europe has
been untouched by the U.S. arms trade.

American weaponry now accounts for about 45 percent of total
world arms transfers, compared with 20 percent in 1970, making
the United States the world's number one arms exporter. Thus the
United States is cynically referred to by some people as "the
world's merchant of death." The Soviet Union ranks second in arms
transfers, exporting about one-half of what the U.S. does. The other
two major suppliers, minor relative to the U.S. and the U.S.S.R. but
still accounting for a combined total of several billion dollars' worth
of arms transfers yearly, åre France and Great Britain. Minor
suppliers are West Germany, Italy, Sweden, Switzerland, Canada,
Israel, China, Poland, and Czechoslovakia (Figure 70).

In addition to the large dollar volume of the United States and
world trade in weaponry, several other important points stand out:

1. The sophistication of arms shipped abroad has drastically
increased (Figures 71 and 73). Such advanced American systems
as Lance surface-to-surface missiles, Hawk surface-to-air missiles,
F–14 supersonic jet fighters, precision-guided munitions, and the
most modern Spruance class destroyers are now being sold to
foreign weapons importers. Occasionally these importing countries
are receiving the newly developed weapons in advance of the
supplier country; the recipient country may even help fund the
development of a particular weapon in the supplier country, as is the
case with Iran importing the U.S. land-based F–18 jet fighter.

2. As sophistication has risen, so have the prices of advanced
weapons systems. A price list of representative weapons on the
world market is given in Figure 71; these costs are not small. In the
case of the Middle East, the practice of "gold-plating" arms
transfers to Arab countries, that is, of raising prices partially by
including expensive optional equipment, has become the norm
rather than the exception; in 1976 this led the Shah of Iran, for
example, to question angrily the four-fold increase in the price of jet

MAJOR SUPPLIERS OF ARMS

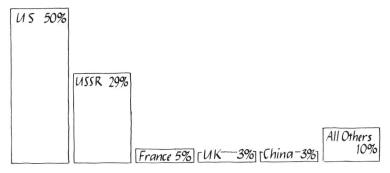

FIG. 70

The United States and the Soviet Union are the preponderant arms suppliers to the world. Of the two-to-three dozen countries which export weapons, the U.S.A. and the U.S.S.R. together account for 75–80 percent of the total volume of arms trade. Most other major weapons suppliers are in Western or Eastern Europe. This small number of major arms producers is in sharp contrast with the large number of arms importers worldwide.

World Military Expenditures and Arms Transfers.

fighters being shipped to Iran from the United States. These price rises, particularly in the Middle East, are an apparent effort to recover U.S. dollars spent for the purchase of oil. In fact, several recent Middle East weapons deals have been barter arrangements—a direct exchange of arms for crude oil.

3. A high percentage of U.S. sales in recent years has gone to the Middle East, an area with obvious potential for conflict. Approximately 63 percent of the foreign military sales over three years—1974–76—has gone to the Middle East, primarily to Iran, Israel, and Saudi Arabia, but to other Arab countries as well. Official government orders of military hardware in the Middle East rose sharply by the early 1970s and have continued at a high level. (See Chapter 10.)

4. The retransfer of American weaponry by recipients, although restricted by legislation, has been unofficially condoned on occasion by U.S. officials. This allows State or Defense Department officials to sidestep restrictions placed on arms transfers to particular countries. It also makes it difficult to determine exactly where

A CATALOG OF ARMS FOR SALE

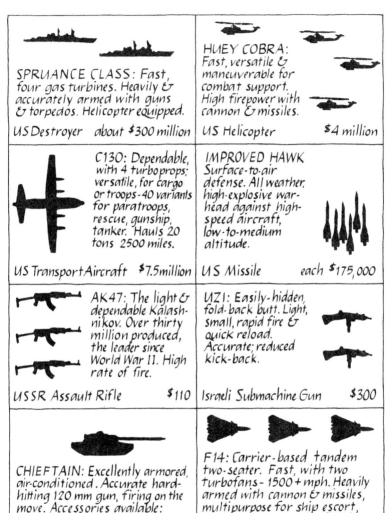

SPRUANCE CLASS: Fast, four gas turbines. Heavily & accurately armed with guns & torpedos. Helicopter equipped.

US Destroyer   about $300 million

HUEY COBRA: Fast, versatile & maneuverable for combat support. High firepower with cannon & missiles.

US Helicopter $4 million

C130: Dependable, with 4 turboprops; versatile, for cargo or troops - 40 variants for paratroops, rescue, gunship, tanker. Hauls 20 tons 2500 miles.

US Transport Aircraft $7.5 million

IMPROVED HAWK Surface-to-air defense. All weather, high-explosive warhead against high-speed aircraft, low-to-medium altitude.

US Missile   each $175,000

AK47: The light & dependable Kalashnikov. Over thirty million produced, the leader since World War II. High rate of fire.

USSR Assault Rifle $110

UZI: Easily-hidden, fold-back butt. Light, small, rapid fire & quick reload. Accurate; reduced kick-back.

Israeli Submachine Gun $300

CHIEFTAIN: Excellently armored, air-conditioned. Accurate hard-hitting 120 mm gun, firing on the move. Accessories available: snorkel & dozerblade.

UK Battle Tank $650,000

F14: Carrier-based tandem two-seater. Fast, with two turbofans - 1500 + mph. Heavily armed with cannon & missiles, multipurpose for ship escort, ground attack & dogfighting.

US Air Superiority
    Jet Fighter $20 million

certain weapons will be used in the future. A weapon shipped to Iran may, for example, eventually end up in Chile or one shipped to Saudi Arabia may be resold to Libya in violation of U.S. agreements.

5. There is no guarantee that recipients of weapons will remain allied to supplier countries. Recent cases are those of Egypt and Somalia turning their backs to the Soviet Union after years of alliance and now obtaining weapons from Western suppliers; similarly, Ethiopia is now securing Soviet weaponry although supplied previously by the United States

As a bizarre consequence of the large volume, high sophistication, and retransfer possibilities of arms sales, as well as the instability of recipient regimes, the U.S. has found itself at times requesting military appropriations to be able to counter the sophisticated weapons it has sold to apparently allied countries. And in several countries now, both U.S. or Western weapons are utilized alongside Soviet or Eastern weapons: for example, both U.S. and Soviet jet fighters are now being flown by Ethiopian pilots against Eritrea.

DIFFERING VIEWPOINTS ON THE ARMS TRADE

We will outline here a few of the arguments used by proponents of the international trade in arms. It is argued that such trade supports allied relationships and thereby promotes United States' international interests and that it maintains stable balances in conflict areas and gains U.S. access to important overseas bases. It returns domestic currency from abroad, for example, Eurodollars and money spent for Middle East oil imports. Moreover, domestic production lines of weapons systems, it is argued, are kept "warm"

FIG. 71

There are a large variety of weapons for sale in the world, from expensive naval vessels and aircraft to inexpensive handheld rifles and machine guns. As weapons have become more sophisticated, so have prices jumped. One new tank may cost upwards of a half million dollars; one new aircraft may cost over $20 million. The representative costs above are estimates; prices vary greatly according to year of purchase, model, contract arrangements, and accessories.

Many public sources, including the JANE'S volumes.

through overseas sales; these sales currently account for about 25 percent of total U.S. production of defense-related hardware among hundreds of industries. Sales abroad also help utilize U.S. defense industrial capacity and thereby provide a reserve capability during emergencies. By increased production through foreign sales, unit costs of weaponry, including costs incurred through research and development, are reduced.

It is also argued that weapons' use abroad in local conflicts affords the United States field tests of its weaponry under wartime conditions. An open arms sales policy does not discriminate between producers and buyers or between haves and have-nots. Since small sovereign countries are free to import the same advanced, conventional weapons systems as those used by the major powers, it is argued that arms transfers help prevent proliferation of indigenous weapons production capabilities or even nuclearization of small powers.

Not all of these arguments are without merit. We believe, however, that the military and economic burdens of the arms trade far outweigh the benefits. Much of the weaponry is shipped to conflict areas; when fighting does break out, as has happened several times of late in the Middle East, the level of conflict is at a much higher level of killing and destructiveness.

Whether or not conflict erupts in these volatile areas, the increasing imports of sophisticated weaponry represent a considerable militarization of the recipient country, its economy, and its ruling elites. Considerable third world defense industries have been developed in the last decade in such countries as Israel and India, often under license arrangements with United States and other arms exporters. Israel, for example, has increased its arms exports, as its defense industry has progressed, from under $60 million yearly in the early 1970s to almost half a billion dollars by the mid-1970s.

Military spending on weapons replaces other more peaceful needs of developing and developed countries. The latest figures on worldwide military spending show that the less developed world is now increasing its expenditures at a faster rate than the developed world. This is especially the case in the Middle East, Africa, and Latin America, shown in Figure 72, where total military spending rose six-fold from $4.4 billion in 1966 to $27.6 billion by 1975 (in 1974 dollars). A large portion of this is due to military arms transfers. At a time of rising demands in the social, economic, and political spheres for developing countries—health, education, population planning, agriculture—military spending for arms imports

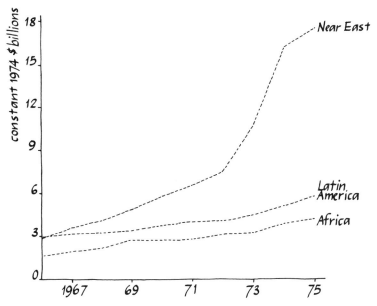

SOME 3RD WORLD MILITARY SPENDING

constant 1974 $ billions

Near East

Latin. America

Africa

1967 69 71 73 75

FIG. 72

Military expenditures have been rising in both current and constant dollars throughout the world over the last decade; this is particularly true for developing areas, where military spending over the last years has begun increasing at a faster rate than in the developed world. The arms trade accounts for much of this increase, especially so in the Near East, where military spending has more than quadrupled in real terms over the last decade. Much of this spending represents foregone investment in health, education, and economic development.

World Military Expenditures and Arms Transfers.

represents large foregone opportunity costs. The developing countries have spent billions of dollars on arms imports from the United States and elsewhere while much of their population remains illiterate, impoverished, and undernourished.

Military aid is also used by the United States in many cases to support repressive regimes. For example, in 1976 the United States aided such countries as South Korea, the Philippines, Indonesia, Thailand, Chile, Argentina, Uruguay, Haiti, Brazil, and Iran; all of

SUPERSONIC AIRPLANE SALES

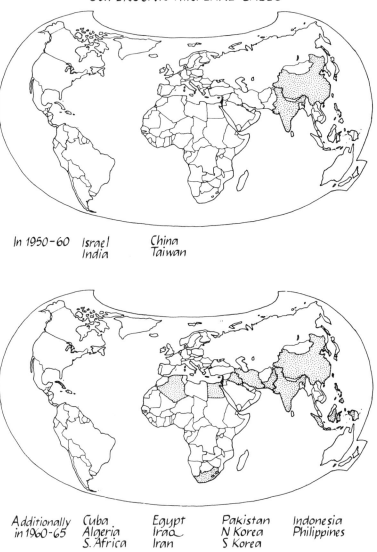

In 1950-60   Israel   China
              India   Taiwan

Additionally Cuba   Egypt   Pakistan   Indonesia
in 1960-65   Algeria   Iraq   N Korea   Philippines
              S. Africa   Iran   S Korea

FIG. 73

Sophisticated and expensive weaponry has spread over the last twenty years to
most parts of the third world. The trade in supersonic aircraft is representative of
this spread. In 1960 four third world countries imported jet aircraft; by 1975,
this number had risen to 43. Many third world countries now appear more

The spread of sophisticated weapons to the third world

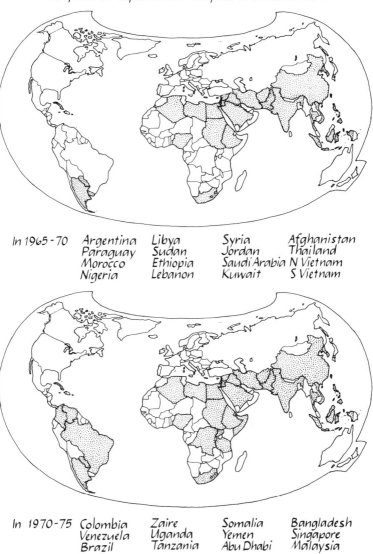

In 1965-70
Argentina
Paraguay
Morocco
Nigeria

Libya
Sudan
Ethiopia
Lebanon

Syria
Jordan
Saudi Arabia
Kuwait

Afghanistan
Thailand
N Vietnam
S Vietnam

In 1970-75
Colombia
Venezuela
Brazil

Zaire
Uganda
Tanzania

Somalia
Yemen
Abu Dhabi

Bangladesh
Singapore
Malaysia

militarily than economically developed. The spread of such arms indicates the rising costs of military imports for the third world, as well as the increase in the destructive potential of local or major conflicts.

SIPRI YEARBOOKS.

these countries have been questioned as to their violation of human rights.

Massive involvement in the arms trade has given the United States a tarnished reputation abroad; the symbol of the American profiteer selling weapons of war is not uncommon today. Such trade tends to involve the United States in conflicts at an early stage; the United States is expected to supply parts for U.S. weapons damaged in any conflict and U.S. personnel are required for training and maintenance programs.

RECOMMENDATIONS: REVERSE THE TREND

In view of all these problems regarding arms trade, we propose that the United States, as the leading weapons seller, take the initial steps to reverse the upward spiral in international weapons transfers. First, while maintaining support of stable power balances in local conflict areas, the United States should take the unilateral step of reducing exports of primarily offensive and highly destructive weaponry and of dual capable systems (systems which may be mated not only with conventional weapons but also with nuclear weapons). At the same time the United States should encourage the other major arms suppliers to follow its example.

The U.S. Defense Department has been reluctant to take this step because it states: "Unilateral weapons embargoes have proved ineffective in controlling arms races." This bald conclusion is based, however, on only one case where the United States temporarily embargoed sales of jet aircraft to Latin America; the recipients eventually switched to alternate suppliers.

What we recommend is serious and purposeful restraint, that is, a minimum reduction by the United States of 50 percent in arms sales. The U.S.A. has indicated a willingness to start toward this direction in recent recommendations from the Department of State to the President and in recent cancellations of some more questionable arms sales such as concussion bombs to Israel. We believe that such a cutback would most certainly reduce the world market in weaponry over the long run. At a minimum it would remove the United States from the very undesirable market in military equipment to a significant extent. In recommending reductions in U.S. arms sales, we do not envision immediate cutbacks of arms transfers within NATO. In particular NATO cases, however, such as the recent Congressional restrictions on

U.S. aid to Turkey, cutbacks may be desirable.

In addition to a cutback in arms sales, the United States should significantly reduce its number of military salespeople and technicians abroad, thereby limiting U.S. involvement in foreign military programs. It should also more closely monitor retransfer of U.S. equipment as it did in blocking the 1977 sale of Israeli aircraft with U.S. jet engines to Ecuador.

The burden of proof must no longer rest with opponents of arms shipments. It should be the duty of U.S. officials to prove that an arms sale is absolutely necessary. This will necessitate closer scrutiny and better coordination of arms sales proposals by federal agencies and the Congress.

The United States should not allow itself to become the sole major supplier to any geographic region: it has come close to becoming such in recent years in the Middle East. We recommend restraining U.S. arms sales by an embargo of particular geographic regions such as Latin America and Africa. Arms sales should be made only where there is an obvious foreign threat to a country's security. Even then, defensive, countermeasure weaponry such as surface-to-air missiles and antitank weapons should be sold rather than offensive weaponry such as surface-to-surface missiles and deep interdiction aircraft.

Defense officials have explained that "recognizing the dangers of growing arms accumulations around the world, the United States has promoted multilateral arms limitation agreements." We see no evidence of this except in a few minor cases in the ineffectual United Nations Conference of the Committee on Disarmament in Geneva. Therefore, in addition to the unilateral cutbacks outlined above, we recommend the United States initiate an international conference of weapons suppliers and recipients to negotiate multilateral limits on the arms trade. This step recognizes the fact that unilateral limits on arms exports by one supplier do not guarantee that other suppliers will refrain from moving into open markets. Considering that the United States has major economic influence on Great Britain, Italy, and Israel and close ties to West Germany and Canada, our influence can be considerable and should be used. Calling for a conference also recognizes that the trade in arms consists of two sides—supply and demand. Suppliers cannot alone stop the arms trade; importing countries, who have in the 1970s shown interest in the United Nations in arms trade limitations, must be included in deliberations. Our goal should be to reduce considerably through immediate multilateral steps the international trade in weaponry, with the eventual goal of eliminating it altogether.

Major related subjects for possible inclusion in conference proceedings should be: standardization of what constitutes an "arms transfer" and institutionalization of a reporting and recording procedure for all such transfers; models for converting military production facilities to civilian-related production; establishment of an international review board of suppliers and recipients to rule on requests for weapons transfers; immediate export limits on particular highly destructive, dangerous, or inhumane types of weaponry; and limits among suppliers and recipients on weapons transfers in particular geographic regions of the world.

It is imperative that immediate, productive steps, both unilateral and multilateral, be taken by suppliers and recipients alike to end the laissez-faire attitude to the burgeoning international trade in weaponry. The United States is in the unique position today to lead such a process. We recommend it seize the initiative to a reduced militarization of the world.

18

MILITARY RETIREMENT
AND COMPENSATION

Over the last decade the U.S. Department of Defense has witnessed a considerable increase in its personnel payroll costs. In 1964 it cost the military about $24 billion for close to four million civilian and military personnel; by 1978 the figure had risen to over $60 billion for about three million personnel. This represents a per capita increase from $6,200 to over $19,000. The most important factor in this cost rise is the "pay comparability principle" underlying the 1970s all-volunteer force. It called for cost-of-living adjustments, military pay increases matched to raises for civilian federal employees, and increased retirement benefits. Two other important elements are inflation and rising costs of retired pay.

MILITARY RETIREMENT

Compensation for retired military personnel alone has risen at the highest rate of all, going from $1.2 billion in 1964 to $9.1 billion in fiscal year 1978. The amount now projected due military retirees is estimated at over $150 billion. Though still a relatively minor portion of the total defense budget of over $100 billion per year, retired pay costs have increased more than seven-fold since the mid-1960s while the defense budget has approximately doubled. In 1964 military retired pay accounted for 2 percent of total defense outlays and 5 percent of the personnel payroll; by fiscal year 1978 these figures had risen to 8 percent and 15 percent, respectively.

This pronounced rise in military retired pay over the last decade has been due not only to inflation, but also the the increase in the number of individuals receiving such benefits. The number of military personnel retired has nearly trebled in size from 400,000 in 1964 to 1.1 million in 1976. This growth reflects both the decline in

PER CAPITA PERSONNEL COSTS

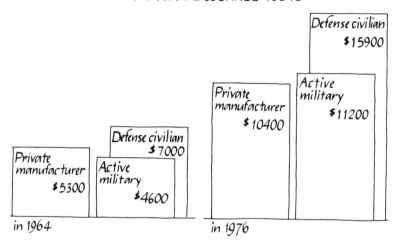

FIG. 74

Active military and defense civilian pay has risen more than private industry pay over the last decade; the former two jumped 143 percent and 127 percent, respectively, between 1964 and 1976, whereas pay in private manufacturing rose 96 percent. Military pay and benefits now account for over 50 percent of annual defense spending.

Departments of Defense and Commerce, as reported in CBO: *Defense Manpower: Compensation Issues for Fiscal Year 1977*, 1976.

U.S. active military forces during the post-Selective Service and post-Vietnam era and the large number of military personnel who entered service during the Korean period. Those persons have recently become eligible for retirement after accumulating 20 years of service—the minimum time required to retire with pension benefits.

Personnel retirement costs now comprise a sizable portion of the U.S. defense budget. It is estimated that they will rise to as high as $20 billion by 1990. Personnel reductions recommended elsewhere in this analysis would cause these projected dollar amounts to rise even further unless the retirement problem is dealt with expeditiously.

We are not alone in recognizing the importance of the increasing burden of retirement costs. In 1971 a Presidentially appointed interagency committee concluded that the military retirement system was costly, inefficient, and unfair. Its recommendations for reform, however, foundered within the Armed Services Committees of Congress. Although review commissions should be more qualified to study the problem than we are, there are several broad points worth outlining here to emphasize the gravity of the problem and the need for reform.

The military retirement system is presently badly organized. Unlike participants in civilian programs, military personnel do not contribute out of their salaries to a retirement fund. Rather, the annuities for retired military personnel are funded through annual Congressional appropriations. The Department of Defense argues that an implicit payment system is in effect since personnel are allegedly paid about 7 percent less than they would be under a regular civilian retirement system. What results, however, if this be the case, is de facto subsidization of retired personnel by those who remain in the service less than 20 years and may never use retirement benefits themselves. The system is therefore far from equitable; it benefits the so-called lifers at the expense of the short-service veteran. We recommend a system in which an individual builds up retirement credit throughout his or her service. Such a system, similar to civilian retirement systems, would over time build up its own equity and not depend on annual Congressional appropriations.

The military retirement system is unfair in other respects as well. Some retirees receive greater compensation than others, depending on the year in which they retired. This is caused by the fact that retirement pay is computed from one's previous terminal basic pay which, after 1968, included "catch-up" pay increases to equate with the civilian economy and items such as housing and food allowances. These inequities must be remedied; one improvement would be to compute an average salary in constant dollars rather than use a terminal year as the base for retirement pay.

The military retirement system is overly costly and inefficient. Eligibility for retirement benefits now starts after 20 years of service. A soldier can collect 50 percent of his or her salary as a pension after 20 years' service or a maximum of 75 percent after 30 years service. The individual who leaves before 20 years' eligibility loses all benefits. This is in stark contrast to civil pension plans where benefits start accruing after five years of work. Moreover, in the military system benefits increase very slowly after those first 20 years. This encourages people to stay in the military until they have

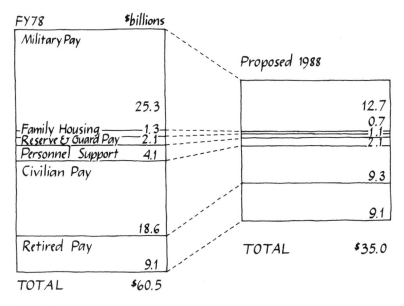

PERSONNEL COSTS

FIG. 75

Defense personnel costs—pay and related items—amounted in fiscal 1978 to over $60 billion. If our recommendations are followed, that is, about a 40 percent reduction in active duty forces, personnel costs might be reduced to about $35 billion (in 1978 dollars) over a ten-year transition period. Retired pay would increase in the interim to cover transition costs, but would eventually settle back down around its present level.

DEFENSE SECRETARY FY 1978.

accumulated precisely 20 years and then retire at an early age. A system must be devised to encourage qualified people to stay in the military beyond their minimum 20 years' service but at the same time to compensate equitably those who retire before 20 years.

The average age at which Army officers retire is forty-six; about 28 percent of military personnel retiring, according to a Congressional study, are in their thirties. Such people who retire early and receive a sizable pension may still take a well-paying nonmilitary federal job. Many people do just that; there are now over 100,000 active federal employees who are retired from the military; about one out of every seven military retirees holds a federal job. We believe that people do not deserve full retirement pay until they actually retire from both military and nonmilitary jobs. If people are not eligible under the civil retirement system—social security—to collect benefits until they are in their sixties, why should the military retirement system be so different? (A civil servant who retires from one job at age forty-five today is not eligible, according to retirement system rules, to collect from the new salary and pension combined more than in the prior salary.)

An equitable scheme is not obvious, however; mid-career pensions in the past were viewed as compensation for long underpaid military service which had developed skills nontransferable to the civilian economy. It is as difficult a problem as the "sportsman syndrome"; there a professional athlete demands a high salary in his or her early years to compensate for the lack of retirement benefits when he or she retires at age forty or so. Today, however, military service is not underpaid and many more military job skills are now transferable to the civilian market.

The news-making military retirements are those of the fifty-year-old generals and admirals, receiving sizable retirement benefits, who then obtain as well high-paying positions with private defense contractors or even with a federal agency. (The latter are double-dipping into the federal pocketbook.) But the majority of cases are not of this type; they are like the colonel or master sergeant who retires at age forty-five with a moderate pension—the average military retired pay is $6,648—but finds himself or herself unqualified for any responsible civilian position. What is needed for these retirees is retraining programs to take care of mid-career job shifts. One possible solution would be for retirees, generals and sergeants alike, to be funded for retraining programs, if need be, of one to four years. Any retirement pension would be deferred until the individual reached the generally accepted retirement age, and then it would be integrated into the social security system.

Military pensions are inflated annually by 4 percent every time the consumer price index increases by 3 percent. This is called the annual "1 percent kicker" meant to account for the lag between inflation and retirement raises. Military pensions have actually outpaced inflation in the 1970s. In the civilian economy, only a minuscule percentage of pensions have such periodic inflation increases. Linking retirement pay increases to the consumer price index may create problems. It may encourage early retirements, since one's projected pension, rising with a high rate of inflation, may increase faster than would one's salary on active duty. A recent Congressional study shows that military personnel retiring after 20 years can expect 32 percent to 44 percent higher income than their pay over the two decades 1955-75; the pension of a civilian worker retiring at age sixty-five would be about one-quarter of the salary during that same period. We recommend decoupling retired military pay from the consumer price index and not overinflating it in the future.

MILITARY COMPENSATION

We realize that reform of the military retirement system is only half of the military pay problem. The whole problem of military compensation has deserved attention in the last few years.

Military personnel costs, which include pay, housing allowances, and numerous other benefits such as travel, commissary privileges, and the like, have risen from $13 billion in fiscal year 1964 to over $27 billion in 1978; this is about 25 percent of the defense budget in both years.

If one includes the two other major personnel costs, civilian pay and retirement benefits, these figures increase to over $22 billion in 1964 and $55 billion in 1978, about 45 percent and 50 percent of the respective defense budgets. It is clear from these figures that personnel compensation makes up a sizable portion of the defense budget. However, the rise in pay rates is clearest when one realizes that the number of personnel decreased 25 percent during this period, from four million down to three million civilian and military employees.

Military personnel costs in 1964 were $4,600 per capita whereas by 1976 they had risen 143 percent to $11,200 (Figure 74). Per capita defense civilian personnel costs in 1964 averaged $7,000 whereas by 1976 they had increased 127 percent to $15,900. By 1978 this figure had risen to $18,500. (Over the period, U.S. Department of Commerce figures for workers in private manufacturing show a 96 percent rise from $5,300 to $10,400. And workers' income is subject to income tax, while much of the military benefits are exempt from any taxation.)

It appears that military and defense civilian pay has risen at a faster rate than has private industry pay over the last decade. This should not be particularly surprising; the increase was an effort to equalize military salaries with civilian salaries in order to develop the present all-volunteer force. There have been definite advantages to this plan. No longer are individuals indirectly taxed by being forced into military service and paid extremely low salaries; a buck private entering the U.S. Army in 1969 before the all-volunteer service received less than $100 per month in salary. Today that same Army private receives a starting salary of over $350 per month. Second, the quality of the soldier has reportedly improved in the all-volunteer force; although the per capita cost is greater, the quality of output may also be better.

Military compensation cannot continue to rise so fast. If it does, the U.S. military will be an extremely expensive force, possibly a force of good soldiers unable to afford proper equipment.

We therefore make the following recommendations for military compensation and personnel policy:

1. Military and defense civilian salaries should not continue to increase at the same rate as in the recent past. They should be fair and equitable, but should not overall surpass the average of the civilian economy. If food, travel, and housing allowances are retained at present levels, then consideration needs to be given to lowering military pay below civilian averages. Given laws of supply and given the personnel reductions recommended elsewhere in this book, we may expect lower levels.

2. The numerous additional benefits, such as commissary and PX privileges, which accrue to military personnel, some good and genuinely needed, should be examined for possible economies. It is doubtful that so many are needed to maintain a volunteer force.

3. "Grade creep" is the process whereby forces age, to become top-heavy with high-ranking officers and noncommissioned officers. Promotion policy must be reviewed and the average grade level periodically cut back. This is done in the civil service and should become common practice in the military. It may present problems of morale, but if it can be done equitably this step can considerably reduce personnel costs.

4. The all-volunteer concept should be reviewed for its continued viability. We recognize the benefits of such a force concept—the reduced imposition upon American youth, the possible higher dedication of standing forces. But the drawbacks are also large—the high cost per capita, the risk of a mercenary character of the force, and the possible nonrepresentativeness of the volunteers.

Our recommendations in other chapters of this analysis amount to reduction of active-duty military personnel by about 40 percent. This must, of course, add to the retired military costs in the short run; but if reform is undertaken, it should not add significantly to pensions in the long run. To illustrate the impact of our recommendations, the $60.5 billion personnel costs for fiscal year 1978 are given in Figure 75.

In the short term, these annual personnel costs would be reduced about 30 percent to $43 billion. This results from cutting active military personnel by 40 percent, thereby reducing the first three categories from $30.7 billion to $18 billion; the fourth category—civilian pay—is reduced from $18.6 billion to $11 billion, but the fifth category—military retired pay—is increased by $5 billion up to $14 billion to account for retraining programs and for more retirees. Over the long term, this last category should settle back to its original amount once the military intake adjusts to a steady smaller number of personnel; this would raise the personnel costs reduction to about 40 percent for a total of $38 billion (in 1978 dollars), compared to the present $60 billion.

PART V
IMPACT OF THE RECOMMENDED CHANGES

19

THE BUDGET: MORE SECURITY AT LESS COST

The estimated cost of the U.S. military forces proposed in the preceding chapers is approximately $73 billion in 1978 dollars. This represents a $47 billion or 39 percent reduction in the current U.S. military budget of $120 billion for fiscal year 1978. As will be explained in the following pages, however, this spending reduction is not expected to be immediate; it can, however, logically be attained several years hence if serious consideration is given to these recommendations.

The military budget of the United States is large, extremely complex, and confusing. Few people understand it or are willing to try to understand it. It takes hundreds of thousands of labor hours, documented pages, and computer printouts annually to produce a 300-page summary of the Department of Defense budget for Congressional and public consumption. The military budget, once $1.5 billion in 1940 at the start of World War II, has now risen from $120 billion to $130 billion in a time of relative world peace. Also, sizable military-related costs such as veterans' affairs and atomic weapons development are not included in the U.S. defense budget. If one includes veterans' affairs and atomic weapons costs, the U.S. military budget is the largest item out of the federal taxpayer's dollar, consuming over 25 percent of federal spending; this percentage would be still larger if one excluded specially funded items, such as social security payments, from the federal budget (see Chapter 5). Looked at on a world scale, the U.S. military budget alone is close to the gross national product of Canada, and is larger than the gross national product of all but eight countries in the world.

Before we discuss fully the budget costs of our recommended policy changes, it will be necessary and helpful to discuss the complex and confusing issues of what one might call "the budget game."

THE BUDGET GAME

"It's Up But It's Down." Depending on how one presents the U.S. military budget, it can be seen to rise or fall or even to fluctuate in annual size. It is commonly presented in five different ways, three of which indicate that it has recently been falling. The first two methods of presentation are to show the budget either in current or in constant dollars. In current dollars, that is, without accounting for the effects of inflation, U.S. military spending has more than doubled over the last 14 years, from $50 billion up to $120 billion. If this is calculated in constant dollars, a method which roughly accounts for inflation, one finds that in 1978 dollars the budget has remained the same, at $120 billion, be it 1964 or 1978; this means that inflation has accounted for the $70 billion rise in current dollars. The military budget did rise to $160 billion in 1968, at the height of the Vietnam War, but declined thereafter. Thus, in current dollars the military budget appears to have risen considerably, but in constant dollars it appears to have remained relatively steady.

Three methods are commonly used to show that U.S. military spending has been going in the other direction, that is, falling. This is done by taking the budget as a percent of something, namely, of the federal budget, of gross national product, or of net public spending during relatively recent times. If one again compares 1964 with 1978, the military budget as a percent of the federal budget has dropped from 43 percent to 25 percent; as part of gross national product, it has fallen from 8 percent to 5 percent; and of net public spending, it has gone down from 29 percent to 17 percent. There are two catches in these methods of comparison, however, which are worth pointing out. First of all, why should the U.S. military—or any federal bureaucracy for that matter—be guaranteed a certain minimum annually of the federal dollar? That is one interpretation of relative percentages in these arguments. It would be better, one would think, to devote to any area the necessary amount of resources regardless of what percentage of something it represents. And second, if the time period of analysis were extended back to 1950—a longer time period is usually a better way of looking at such figures—one sees that these three percentages have not fallen at all! On the contrary, they have actually crept slightly upward over the three decades.

"What's the Right Number?" In addition to being wary over statements about increasing or decreasing budgets, one must be careful to specify which numbers one is using when analyzing the defense budget. Depending on the level of analysis, the military budget figure for any given year can vary more than 10 percent. It does this in four ways.

First, as mentioned earlier, a budget figure varies when one compares current with constant dollars. The fiscal year 1978 budget, in current dollars, is $120 billion. In constant 1977 dollars, this would be $113 billion. In other words, $7 billion over the past year was accounted for by inflation.

Second, the Department of Defense presents three related figures for its budget every year: "total obligational authority," "budget authority," and "outlays." The 1978 fiscal estimates for these are from $110 billion to $120 billion. *Total obligational authority*, the most commonly used figure, represents all monies authorized for the budget to be spent through fiscal 1978. *Budget authority* is all monies authorized in 1978 to be spent both in 1978 and in all future years. And *outlays* are the funds that will actually be spent only within the one fiscal year, here FY 1978 (October 1, 1977 to September 30, 1978).

Third, to make matters still more complicated, the Defense Department annually presents another variation on the budget game—a "total budget" and a "baseline budget." The total budget for 1978 is $120 billion; however, the baseline budget is only $110 billion. The latter does not account for items seen as not directly related to national security; it therefore excludes military retired pay and military assistance.

Fourth, the official Defense Department budget of $120 billion does not include all military costs; it omits the costs for veterans' affairs and for nuclear weapons development and production. These latter costs, such as those for development and potential procurement of the so-called neutron bomb, are included in the budget of the Department of Energy. If these costs are counted, total fiscal year 1978 military costs increase by 16 percent to $140 billion.

"What Does It All Mean?" If all this budget analysis and jargon sounds confusing, that is because it is confusing. It is a difficult subject to comprehend, let alone analyze, and has therefore gone relatively unquestioned for far too long a time. What has resulted from public and Congressional lack of understanding are two assumptions: (1) more military spending means better security; and (2) military spending must be increased yearly to keep pace with security demands. These assumptions, however, are not necessarily true. Does increased health spending, for example, mean necessarily better health care? Must welfare or unemployment spending necessarily increase each year? The answer to both these questions is no.

Similarly, increased military spending does not necessarily improve a country's national security; dollars do not directly equate with defense. In fact, more dollars may even mean less defense. This has been illustrated quite clearly in the strategic nuclear field, where the Defense Department has spent billions of dollars annually to improve and expand strategic forces. The U.S. nuclear arsenal has expanded from a half dozen bombs in the 1940s to over 30,000 nuclear weapons in the 1970s. Has this lessened the chances for a nuclear war? No; on the contrary, it represents a diminution in security from this enormous increase in damage expected from any nuclear exchange.

The American military has been accustomed to a "blank check" appropriations process in Congress. The American public can no longer afford this luxury since it leads to inflation, unemployment, and related economic ills. It is a process that has produced more antagonism than security, more enemies than friends, inefficiency rather than economy, and planning based on bureaucratic rationale. For example, the three military services—Army, Navy, and Air Force—have each maintained close to one-third of the combined funding over the last decades. With such very different equipment, personnel, and military applications, this equal allocation of monies appears more bureaucratic than rational or efficient.

With increasingly limited resources worldwide and with rising demands in every sector of social planning, the United States cannot afford not to examine its military budget closely for ways of economizing. All bureaucracies, especially those which spend the most—namely the U.S. military—need to have the burden of proof placed upon them. This is what we have sought to do; we have investigated the security needs of the United States and have stated the means for achieving them at less risk and at less cost.

A COMPREHENSIVE APPROACH: THE PROPOSED BUDGET

This book has presented a comprehensive analysis of military policy and planning for the United States. Congress, in its responsibility to hold the federal purse strings, has taken a piecemeal approach to the military budget every year, adding or subtracting bits and pieces at the edges. This is not wholly its own fault, and it has been improving its methods through such offices as the Congressional Budget Office, the Office of Technology Assessment, and the General Accounting Office. But what has been glaringly absent is a comprehensive look at military spending priorities. We aim to fill that void. President Jimmy Carter addressed this problem by saying: "Most of the controversial issues that are not routinely well-addressed can only respond to a comprehensive approach; incremental efforts to make basic changes are often foredoomed to failure."

This chapter was one of the final ones written in the book. We started our analysis with a lengthy examination of U.S. national security needs, foreign policy goals, and military capabilities. Only at the end did we arrive at budgeting our proposed force changes. In other words, we did not set for ourselves a goal of cutting the military budget by an arbitrary percentage; we rather outlined our recommended forces, and then proceeded to determine what the resultant costs would be.

Using the fiscal year 1978 military budget—total obligational authority in current dollars—of $120 billion (the latest figures available in detail for this book), we have determined through our analysis that enhanced security can be purchased with a 1978-equivalent budget of $73 billion. This represents a reduction of 39 percent overall—a savings to the federal government and to the taxpayer of over $47 billion. If one factors out two items—retired pay and foreign military assistance—which less directly affect U.S. national security, the proposed baseline budget represents a reduction of 45 percent.

The present and proposed military budgets of the United States are presented below in two breakdowns, both derived according to Department of Defense categories as they are submitted annually to Congress. The two breakdowns represent different subdivisions of the same budget.

1. The first breakdown, in Figure 76A, is according to "military program." It breaks costs down into strategic forces, general-purpose forces, and eight related activities such as intelligence, training, supply, administration, and retired costs.

2. The second breakdown of the U.S. military budget, shown in Figure 76B, is according to "appropriation title"; this shows the costs for military personnel (pay), operation and

THE MILITARY BUDGET:

by MILITARY PROGRAM

FY78 $billions

Strategic Forces	
	10.8
General Purpose Forces	
	42.2
Intelligence & Communication	
	8.1
Airlift & Sealift	1.6
Guard & Reserve	
	6.9
Research & Development	
	10.8
Central Supply & Maintenance	
	12.0
Training, Medical & Personnel	
	15.3
Retired Pay	
	9.1
Administration	2.3
Support of other Nations	1.3

TOTAL $120.4

Proposed 1980 s

1.1

24.3

5.1
1.4

4.3

4.1

7.9

10.2

12.9
1.5
0.4

TOTAL $73.2

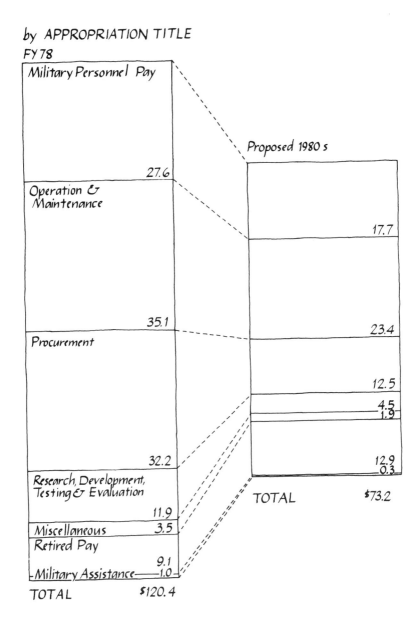

by APPROPRIATION TITLE

FY 78

| Military Personnel Pay | |
| | 27.6 |

Operation & Maintenance

35.1

Procurement

32.2

Research, Development, Testing & Evaluation

11.9

Miscellaneous — 3.5

Retired Pay

9.1

Military Assistance — 1.0

TOTAL $120.4

Proposed 1980 s

17.7

23.4

12.5

4.5

1.9

12.9

0.3

TOTAL $73.2

maintenance, procurement, research and development, and six additional smaller categories such as family housing.

If one looks at the budget by military program, illustrated in Figure 76A, it will be noted that the largest reduction—90 percent—comes in the first category, "strategic forces." This is a result of the recommendation, explained in Chapters 6, 7, and 8, to eliminate several large operations and extremely expensive new procurement programs of the so-called strategic triad of forces. The proposed budget covers support of an ample undersea deterrent force of Poseidon nuclear submarines and of a small force of air defense interceptors. It mainly eliminates the land- and air-based strategic forces presently deployed—the land intercontinental ballistic missiles and the long-range bombers—and eliminates such new, unnecessary, and potentially dangerous systems as the B-1 bomber (cut off by President Carter in mid-1977), Missile-X, the cruise missile, and the Trident submarine. These cuts in the strategic forces budget, less dramatic dollar-wise in 1978 than the expensive general-purpose forces, nevertheless represent enormous future savings in procurement and deployment of these strategic systems.

Because of our cutback in development and procurement of new systems, the largest cuts in other military programs (excluding the "support of other nations" category) come in research and development—62 percent. The remaining categories are reduced much less—anywhere from 40 percent to 10 percent—due to continued maintenance of a strong conventional force. The major reductions in the Army and the Marines come from a decline in active division-equivalents from over 20 to about 12; of these, the Marines themselves are reduced from three divisions to one division. The major Navy cutback is in the reduction of aircraft carriers from 13 to

---

FIG. 76

The U.S. military budget—about $120 billion in fiscal 1978—is annually presented to Congress in two break-downs: by military program and by appropriation title. If our recommendations are followed, it should be possible over the next few years to reduce the budget by about 40 percent. (This estimate includes an interim rise in retired pay to cover the military-to-civilian transition period. If one excludes retired pay, the reduction comes closer to 45 percent—a potential saving of almost $50 billion.)

DEFENSE SECRETARY FY 1978, especially appendix C, pp. C-11 to C-14.

three; the carrier aircraft are largely retained in the active inventory for land-based use. The surface fleet gradually diminishes through retirement without major additional new shipbuilding. The Air Force tactical fighters, interceptors, and other aircraft are retained at present levels. Supporting services such as supply, intelligence, training, and administration are reduced less than the strategic and general-purpose forces due to our calculation that these support activities will decline more slowly and in less than a direct proportion to the military forces. (It is possible that supporting services would in the event decline even more than combat reductions due to inefficiencies in the present structure.)

In both budget breakdowns, the retired pay item is the sole category to increase. The 42 percent rise reflects the considerable number of reductions in personnel in the various services. The figure includes severance pay and retraining costs. Once the new military policies here recommended are fully implemented, these retired costs will rise at a much slower rate.

If one examines the budget from the second perspective, appropriation title, a different set of dollar amounts is found. There are four major categories accounting for the majority of funds in this breakdown: "military personnel," "operations and maintenance," "procurement," and "research, development, testing, and evaluation." The first two categories are reduced by about one-third, whereas the largest cuts come in procurement and research. We sought to eliminate research and procurement of weapons systems which did not fit in with our present recommendations; this entailed eliminating or cutting back unnecessary, redundant, and dangerous weapons projects. Lest the reader think that we wielded the knife arbitrarily, we point out that some systems were fully eliminated, others only partially, and some retained at present or higher levels of funding. The procurement and research and development decisions were therefore examined at length, item by item, to determine what weapons and research was necessary for the retained military forces. We retained, for example, current levels of tanks, tactical aircraft, and surface ships; procurement of these items will, of course, decrease over time as inventories are filled and as the number of active divisions drops.

Just as the above procurement decisions were mixed in determining the proposed budget, so also were the research and development decisions, of which we give samples in the notes.

We make no presumptions of full exactness for our recommended budget; this is impossible to accomplish, even if one had the advantage of a thousand heads and hands as does the Department of Defense. There is uncertainty and error in the official Defense Department budget, as one realizes if one compares official budget forecasts from year to year. For example, the 1978 budget was predicted in 1976 to be $120 billion; by 1977, estimates had risen as high as $130 billion. The budget finally exited from the Pentagon as $123 billion, and was subsequently changed to about $120 billion by the incoming Democratic administration.

The figures in the budgets above have been painstakingly calculated by various methods. Wide use was made of unclassified budget documents from the Department of Defense. In seeking to identify and estimate as much of the budget as possible, we broke it down by services—Army, Navy, Air Force, Marines, and various Reserves—into a large matrix relating military programs to appropriation titles. In other words, the categories in Figure 76A were aligned on a vertical axis, and those in Figure 76B on a horizontal axis. This has never been publicly done before, to our knowledge: It allowed us to estimate more accurately our own proposed budget.

Information concerning the U.S. Defense budget is voluminous, but still it has gaps. The first lack is that some categories of budget items are vaguely defined in official documentation as "special activities" or "support activities," while others lack even any such vague description. One item which falls into this category is covert intelligence operations. One might recall an incident, related to this budget problem, told of Robert McNamara when he was Secretary of Defense in the 1960s. Upon assuming office, McNamara asked his military aides what purpose the hundred-odd ratio antennas on top of the Pentagon roof served; when the aides offered no adequate explanations, McNamara had all the antenna wires disconnected. As he had surmised, there was little response from the Pentagon; apparently few persons missed the cutoff of antennas and, for those who did complain, the antennas were reconnected. We have followed this McNamara strategy in budgeting; when these minor items could not be sufficiently identified, their funding was partially or fully reduced. (The total sums concerned are at the level of a low percent of the full budget.)

A second problem in budget costing comes from items which, although identified in the aggregate, are nevertheless extremely difficult to cost because of the diverse support nature of the activities. Items in this category are administration, supply and maintenance, and training costs. We have reduced these relatively less than the forces they support, on a determination that a minimum military infrastructure—depots, hospitals, headquarters, schools, and so on—is necessary, even with reduced forces. Thus the programs in Figure 76A, "Central Supply and Maintenance" and "Training, Medical and Personnel," decline only 34 percent and 33 percent, respectively, while the forces they support, "Strategic Forces" and "General-Purpose Forces," drop by 90 percent and 42 percent, respectively. One may find over time that these support costs can be reduced further; for example, training costs may go down faster with fewer personnel to train and more troops remaining in service longer. One should remember that the goals in such costing and planning are two: to improve and to maintain high efficiency and to build a superior fighting force, although reduced in numbers.

A third problem in any complex budget analysis and planned reductions is timing. How quickly can a particular budget be increased or decreased without unnecessary repercussions in other sectors of the economy? The budgetary figures used in this study— the fiscal year 1978—were the most recent available. The reduction from $120 billion to $73 billion is not intended, however, to indicate an immediate cut. The proposed budget reduction, as is apparent throughout the book, is intended to take place through gradual decreases—a winding-down effect. Some items in the defense budget such as personnel can be cut relatively quickly while other items, such as major ship-building programs in which ship keels and hulls are already laid, will take longer to reduce. The 39-percent reduction would in all probability not be fully realized until 5 or 10 years hence.

No proposals have been made in this chapter concerning what alternatives exist for the many tens of billions of dollars in savings from the preceding budget recommendations. The point to be emphasized is that the savings are major: $47 billion in 1978 dollars. The savings represent over 10 percent of the annual federal budget and more than the gross national product of many countries in the world. It is more than is spent annually by the U.S. government on social investment and services. To place such an enormous sum in clearer perspective, one can picture it in terms of the average taxpayer; if the money were rebated to the American public, it would save the typical four-person family in the United States one-third of their federal income taxes—or over $1,000. Whether these savings are returned to taxpayers in the form of reduced taxes or reinvested in other federal programs, the possibilities for alleviating pressing domestic problems in the United States are infinite. Certainly employment would be increased, as discussed in Chapter 20. On a worldwide scale, it has been estimated that a minimum investment of one-third of such an amount would considerably lessen, and in some cases eliminate, infectious diseases, malnutrition, infant mortality, illiteracy, lack of housing, diseased water supplies, and would ultimately improve population planning and economic distribution.

20

IMPACT ON JOBS
AND THE ECONOMY

THE EFFECT OF THE PRESENT MILITARY
SPENDING

The effect of the present level of military expenditure on the American economy has been virtually ignored in most discussions about military expenditure cuts. From 1950 to 1975, approximately $1,500 billion was spent by the United States on the military, without recalculating the dollars for inflation. The impact which might have been made on the civilian economy by alternative allocations of this sum is enormous. Had military appropriations been invested instead year by year in the American economy, at a normal 8 percent return in dividends and capital growth, its current value would exceed $5,000 billion. A sum of $5,000 billion is a gigantic figure. It represents a three-year total of the gross national product of the United States: it is seven times the value of all corporate stock listed on the New York and American Stock Exchanges; it is perhaps equivalent to the value of all the land, buildings, and assets in the United States. It has been estimated that $50,000 is the investment in equipment or education necessary to create one new job; by that crude measure, the $5,000 billion could gradually have created something closer to 100 million civilian jobs instead of sustaining only 5 million to 10 million jobs in the military sector (Figure 77). If instead of spending money on the military the United States had concentrated on improving the productivity of people and equipment, America could have entered an economic age of abundance for all. In actual fact, American per capita real income, which had risen historically at about 2 percent each year during the past 100 years, showed a perceptible drop in the decade 1965–75.

In problem area after problem area of contemporary America, a lack of money blocks solutions. Deteriorated housing, worn out rail and subway networks, industrial obsolescence, pollution, exhausted soil and forest, all require heavy, long-term capital in the hundreds of billions of dollars. Raising educational standards, preventive health care, and job retraining all demand millions of people working for years to create new solutions.

The money, the talent, and the personnel to solve these problems can and should be returned to the civilian sector from the military. Granted, military spending provides some jobs. At present, approximately seven million Americans are paid for producing and operating tanks, ships, planes, and rockets, which at best will never be used, and at worst will destroy a great part of humanity. These military items do not add to the pool of goods and services which satisfy ordinary needs and desires, nor do they have the productive effects of schools, hospitals, or civilian factories—investments which help to create further goods and services. Government military spending which creates demand without either creating supply to satisfy that demand, or taxing to remove the demand, feeds an inflation which cannot be cured even by recession.

A GRADUAL WINDING DOWN

Our new military budget will cost $70 billion per year, after the new equilibrium level has been reached, compared to the government fiscal year 1978 budget of $120 billion. We assume that at the end of a 5- to 10-year period our proposal would result in about three million fewer persons engaged in military activity directly and indirectly. Our proposed changes toward lower military budgets will probably take place gradually over this period, in part because we have proposed gradual change, and in part because it will take time for the proposals themselves to win acceptance. It is likely, then, that the job shrinkage would also take place gradually, eliminating about 400,000 positions a year. Assuming a 4 percent annual turnover on seven million military-oriented jobs, much of the shrinkage would be taken up by people normally changing jobs or retiring, and to aid this, we recommend a no-fire no-hire policy be instituted to the extent possible in the military services and occupations.

In each year probably less than 200,000 persons on the average would actually lose their present jobs through layoffs and enter the civilian labor market. In some years there may be of course

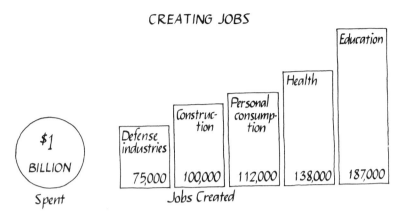

FIG. 77

Millions now idle need to enter the work force; they can be productively employed in health, education, and the home building industry, and trained at work in more skills. Defense industries cannot use them at all, because defense requires highly skilled, highly-paid professionals and technicians. Are military-related occupations used as a kind of white-collar WPA for excess professionals and technicians?

Bureau of Labor Statistics study cited in *Priorities*, 1976.

considerably fewer layoffs, or more, depending on which particular proposals are adopted in any given year and how they affect the level of uniformed or civilian personnel. We will next consider the essential task of ensuring that these persons not suffer undue financial burdens, and that their skills and interests are used in productive work.

THE PERSONAL TRANSITION FROM MILITARY TO CIVILIAN SKILLS

Compared to the rapid ups and downs in employment experienced in the course of American business, and particularly compared to ending World War II war production in 1945 and also demobilizing eight million soldiers in a single year, the problem of helping 200,000 persons a year find work is not beyond solution. We present five components of this solution, emphasizing that this represents but a beginning sketch of the situation.

1. Termination pay should be provided for those laid off from military-connected occupations. A rate equal to one year's pay per ten years' work in military service, the defense industries, or research would provide more than comparable layoff benefits offered in virtually any civilian work.

2. An inventory of personnel skills and interests should be taken and matched with a national job requirements inventory. Government contracts to personnel firms for this phase of conversion should aid considerably in the process. The cooperation of corporations, unions, and state and local governments will have to be provided, and if necessary required by federal law, so that veterans and military work force personnel not be barred from suitable jobs by seniority or craft requirements appropriate to civilian employees, perhaps by equating military service with seniority or experience in civilian employment.

3. While many uniformed and civilian employees will be leaving military work that has direct civilian analogues—driving trucks, purchasing equipment, working in hospitals—some military work will have no civilian counterpart. A new G.I. Bill of Rights should be extended for departing members of the armed services, Department of Defense personnel, and civilian employees in military-related industry, so that useful skills can be acquired or supplemented when necessary. Specialized training programs at school, college, and graduate levels will be necessary. Fortunately the nation has already adequate educational capacity and experience to provide the training. Indeed with the drop in youthful population coinciding with an increasing surplus of teachers, the times are ideal for a large-scale expansion in adult, career-oriented education.

4. A substantial percentage of those employed in the military work force will not change their jobs at all. Only the contracts on which their employment depends will shift from military to civilian applications. Thus steel workers will produce less for tanks and more for buses, but will remain at the same work place. This aspect of the transition will be discussed in the next section of the chapter.

We estimate that from $3 billion to $4 billion per year will represent the direct costs of personnel severance pay and educational benefits during the transition period, and we have budgeted for this item under the category of "pensions" (see Chapters 18 and 19).

5. Substantial additional economic assistance should also be considered. While almost every region and state will increase net employment as a result of a military spending reduction and

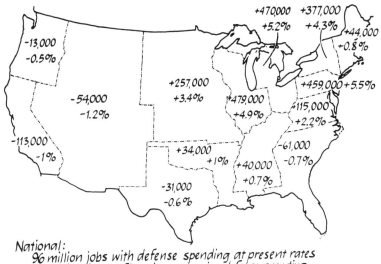

THE JOB MARKET FORECAST

-13,000
-0.5%

+470,000 +377,000
+5.2% +4.3% +44,000
+0.8%

-54,000
-1.2%

+257,000
+3.4%

+479,000
+4.9%

+459,000 +5.5%
+115,000
+2.2%

-113,000
-1%

+34,000
+1%

-61,000
-0.7%

+40,000
+0.7%

-31,000
-0.6%

National:
96 million jobs with defense spending at present rates
98 million jobs after decreases in defense spending

FIG. 78

Studies indicate that 2 million more people will have jobs, largely in the older industrial states of the North, if the military budget is cut by 30 percent. Perhaps the decline of northern U.S. cities and economies has in part been caused by the high level of military spending over the period from 1950 to the present.

R.H. Bezdek, *Journal of Regional Science*, v. 15, n. 2, 1975.

increased civilian production (see Figure 78), particular localities would be sharply hurt, in the absence of compensating federal action. There is no economic justification for laying extraordinary burdens on the citizens of localities which have been hosts to military installations or industries. A sliding scale of federal loans or grants, depending on the degree of local dependence on military activity, should be considered to pay local indebtedness and expenses incurred to meet needs caused by military activity. The situation is complex and lies outside the scope of this book, but a relevant model exists in federal aid for areas struck by natural disasters.

As to the effect of military cuts on corporate profits and structure, we believe that the risk-taking nature of the American economic system, the gradual pace of the reduction, and the habit of corporate executives to plan ahead should minimize the need for large-scale federal intervention for corporate benefit, as distinct from the measures we have proposed for the individuals and localities affected. As one example current in 1977, the officials of Rockwell International, after the multibillion-dollar B-1 bomber production was canceled, informed their stockholders that the adverse effects on their diversified corporation—the prime contractor for the B-1 —would be minimal. Not all military producers would be so resilient. Yet much of the capital for military production has been furnished by the government, and military sales are not particularly profitable or necessary to most large corporations. We have already recommended, in Chapter 15, creating an industrial reserve system for rapid standby military production, and that part of the physical equipment which is necessary for war production can be mothballed in the industrial reserve.

MORE JOBS, REAL GOODS, AND REAL SERVICES

Several economic studies have indicated that, far from creating jobs, military expenditures increase unemployment compared with equal levels of spending for civilian goods and services. The Bureau of Labor Statistics, for example, recently analyzed the effect on employment of shifting $1 billion from military spending to a variety of nonmilitary expenditures (Figure 77). If similar proportions hold true for the scale proposed in our book, one and one-half million net additional jobs will be added to the work force, for a deep cut in unemployment. The capital-intensive nature of modern military expenditures requires a relatively smaller number of people—highly skilled, scarce, and therefore expensive—than do comparable civilian expenditures, which are on average able to use and train more people of average skill. Even more important, since none of the military-employed people are producing goods and services wanted by themselves or anyone else, their own demand is inflation producing, competing for the real goods and services produced by everyone else. In contrast, every single former employee of the military service who finds civilian work, and every new employee employed with funds formerly dedicated to the military, will be adding in some way to real goods and real services demanded by the population as a whole.

Other studies strengthen the case argued here on the basis of the work of the Bureau of Labor Statistics. In one, a 30 percent overall cut in military expenditures was figured, and then its effect calculated on employment throughout the American economy, divided into 80 industries and into 14 geographic regions. Figure 78 reproduces the results in the regions.

The reallocation of defense expenditures to domestic public programs would increase national employment two percent or by two million persons, assuming the changes had taken place by 1980. The regions benefiting most are in the old industrial North and the Middle West (Figure 78). These regions add from $3\frac{1}{2}$ percent to $5\frac{1}{2}$ percent to their total employment, for a net total of two million additional jobs. In the rest of the country employment breaks even, with the two regions most affected negatively, California and the Mountain region, losing 1 percent, or 167,000 net jobs.

There are other models which could be used to examine the economic effects of lower military budgets. As examples only of other assumptions for such economic studies, the $50 billion savings from lowered military expenditures could be directly allocated to individuals either in the form of individual income tax cuts, or in a lowered level of required social security contributions, or in the form of a negative income tax. Obviously these different methods would spread the benefits among differing income groups. Revenues saved could also be shared with state and local governments, or even used to eliminate massive federal deficits.

What is important to realize is that there has been to date little study of *any* of the possible alternatives to massive federal spending on the military. There have been isolated studies of the effects and possibilities of disarmament, as well as studies of particular problems of industrial reconversion. We believe that these studies were limited in their influence by two factors. First, the American political climate of 1950–70 was preoccupied with the cold war and the war in Vietnam. The climate of opinion has changed. Second, there has been no serious examination of the overall military expenditures actually necessary for the military defense of the United States and for its foreign policies. This book begins the inquiry.

21

CONCLUSION

ARMS AND POLICY

The reader has heard all too much of guided missiles and satellites, or reservists and procurement. Clearly there is a surfeit of strange detail in all this account of weapons and the organizations that conceive, produce, and use them. The complexity is understandable enough; after all, this is the largest industry in the world, and it partakes of the most elaborate technology, embedded in a history of the highest antiquity. It could not be other than complex.

But our purpose is not the understanding of arms, or even of war. Rather we aim at setting a more rational policy for the military, in the first instance here in our own country, then perhaps by example and influence elsewhere. Of course policy flows from many other founts than the military. Arms and armies are its mere instruments; policy arises from aims, tradition, economics, cultural differences and change, and much more besides. The whole fabric of relationships among nation-states, and within single states, is the cloth of policy. We have chosen to treat a single side of it. Why have we written so military a book when we might have spent our time on the goals of peace?

First of all, we believe the book is a needed and a novel one. There is a large expert community of people who study and write on military affairs, yet there are very few treatments of the present military policy of the U.S.A. as a whole among the many specialists who look at one or another glittering facet. Yet it is the whole, however intricate its growth, that must be seen. Perhaps the best measure we have for it is the cost of the human labor and resources it embodies, best measured as matters now stand by the military budget. We sought to examine that budget in some detail to see the system behind it, a system conforming to many features of the physical and political world of our times. We might have argued for change in American perceptions of the world, or some new diplomatic stance. We have, rather, fixed our attention on the forces and their economic support as the most objective and the most concrete measure of effective change. We do not claim that here is the root of war or of safety; we do claim that wars are fought by armed men, and in the end it is the use of such forces that determine war or peace.

Second, there is something new in our time. We gave figures to demonstrate how large, important, and dynamically growing is the art and practice of war over all the world. If war is meant to carry out policy, by now policy is much affected in return by the sheer importance of the machines of war, and their organized servants. Each year the weapons reach farther, move faster, explode with incredible potential damage to wider areas. But the surface of the earth grows not an acre larger. World War I touched the imaginations of all because it was seen as worldwide. But it was mostly European. World War II transcended it; there was no Asian front in 1914–18 to compare with the Pacific and Indian Ocean war of 1940–45. Nevertheless, World War II left the southern hemisphere mainly at peace. Nowadays the secondary fallout from a major nuclear war will invade the atmosphere of countries everywhere, neutral, allied, or full belligerents, last, to be sure, the lands below the equator. Mature technology has given its power to war. We cannot neglect that. Even more strikingly, the dynamics of technology have entered warfare; the changes in weapons are more rapid than ever before, less restrained. Weapons seem almost to spawn autonomously in their breathtaking power and variety. Warheads, each capable of killing a whole city, can fly in thousands at a radio command halfway around the world inside an hour. This physical transformation is the outward and visible sign of the inward institutions of war today; we believe that even fundamental policy cannot be argued until these mere instruments have been reckoned with in the mind, not merely statistically, as they are, but as they will come to be if we do not change.

We do not claim the story we tell is the whole story. Behind the weapons are a multitude of issues at every level which can influence history. In the end those issues determine, for weapons do not yet make war of themselves. But by our attention to military matters, we believe we view the chain of events at one particular link, a key link too long overlooked by those who criticize and plan for national policy. At the same time, we have tried to take into account the relevant political and geographical context surrounding the weapons and the troops, for too narrow a look at arms themselves would lack reason. Before the policy we foresee can become reality, much must happen far outside the centers to which we pay attention, permeating the entire body politic. It will be for others to treat these many changes, in diplomacy, statecraft, in the regions of economic gain and loss, in psychology, and in the formation of the entire climate of opinion. Arms are a major possession, though they are not the focus of the state. But they are *our* focus, and we have submitted for the reader's attention what they mean and how they might be changed for the United States.

THE INVESTIGATIONS

Four sorts of material make up the volume above. One detailed body of matter gives a description, with some evaluation, of the current structure of the American military forces and their funding. This has been derived mainly from fairly complete official sources; while there are both errors and officially concealed structure—hidden apparently for reasons of security—we believe the facts are reliable and the interpretations on the whole sound. Second, less concretely, there is some account of the expected use and effect of modern weapons and military organizations, and indeed some projections to the relatively near future. In their very nature, these are less based on simple fact, and represent what we believe are careful and sensible extrapolations and distillations from experience and the world literature of such commentary. We cannot claim that our crystal ball is free of flaws, any more than is that of the Pentagon. But we have made generous allowance for uncertainty and error; within the limits a sensible reader can expect, this forecast is quite adequate for guidance. Third, we have used official U.S. documents and the standard references to estimate, in much less detail, the size and the nature of the forces against which ours might potentially become engaged. Moreover, the fundamental strategic geography and economics of the chief potential regions of conflict have also been put forward, in considerable brevity. Strategic judgments, even hedged as we have tried to present them, are notoriously artistic. One witty writer remarked: "Before every war, military science looks like astronomy. After each war, it looks more like astrology." Still, we have given our reasons for the estimates we make in terms of what the conditions of conflict might be.

The fourth and final segment of our study includes our recommendations for the forces the U.S.A. ought to retain and the goals we foresee in each of the possible regions of conflict. Here we were as forthright as we could be; we indicated a rough decision at more or less every important point. There is no doubt that here and there we will turn out to be wrong, out of ignorance or defective reasoning. But we put this scheme forward, not as beyond amendment or criticism, but as a coherent and reasoned approach to the national defense, an approach whose novelty of assumption and conclusion we believe few will dispute. This book is not the last word in a long argument, but one of the first. We welcome informed correction of fact and reasoned difference of conclusion. Until a substantial body of opinion has formed around the idea that a safe and defensible military policy is not the same thing as just saying *more*, our hopes will be without issue.

Not everything can be pinned to detailed judgments of what is workable in each specific circumstance. So overbrittle a position would not in itself be as reasonable as we claim, even if each decision were based on good points. We therefore curbed our judgments in two ways which are more formal than the discussion of realities and projections. First, we recognize that in complex human affairs it is not always prudent to change too rapidly, even when one is convinced of the need and desirability of change. For this reason we frequently adjusted our expectations to a slower rate of change, one that gives a chance for all the complex reactions to begin to manifest themselves. Such a scheme gives safety against error in a most understandable way. Second, we looked for the paths of retreat from our own decisions, should they turn out wrong. An irreversible change can be made only upon very firm grounds; but a change which contains a built-in scheme for returning to an earlier state of affairs can be begun with only plausible expectations of success. Many of our recommendations will be seen to contain this possible reversal as part of gradual implementation over the run of some 5 to 10 years; this is the typical—though not the universal—time scale of our proposals.

AN OVERVIEW OF PRESENT AMERICAN FORCES AND THEIR RATIONALE

As the speeches often go, the U.S. military forces are demonstrably "second to none" all across the board. Only in military manpower

A PROPOSAL FOR THE MILITARY FORCES

STRATEGIC FORCES	Present			Proposed 1980s
Offensive:				
Missile submarines	41			31
Land-based missiles	1050			100
Bombers	380			0
Their tankers	480			150
Defensive:				
Aircraft	148			148

GENERAL PURPOSE FORCES

Aircraft			
Air Force	2300		2300
Navy	1550		1350
Marines	450		450

Troops:			
Army heavy divisions	10		8⅓
Army light divisions	7⅓		3
Marine light divisions	3		1

Warships			
Aircraft carriers	13		3
Major surface ships	162		125
Attack & patrol subs	78		70
Amphibious warfare ships	63		10

Airlift & Sealift			
Aircraft	550		550
Ships	50		50

FIG. 79

How the military forces of the United States look overall before and after our recommended changes.

Afterwards:

i) Strategic forces: The submarine deterrent is retained, the least vulnerable could that proposition be questioned, for countries of larger population can and do arm more men. In conventional land forces, the U.S.S.R. and its allies meet U.S. forces and U.S. allies on somewhat even terms in Europe. In the air over the battlefield, the U.S. Air Force leads all others, and the U.S. Navy with its carriers has no rival. Indeed, only the Soviet Navy can be said to have a naval strength in the wide oceans which is anywhere similar to the U.S. naval capabilities. In the nuclear domain, the United States has no superior again; the U.S.S.R. has strategic forces which are broadly comparable but no other countries can so claim.

But of all these U.S. forces, a sober view must conclude that well less than one tenth, perhaps only a few percent, are forces actually in direct defense of the United States. This fraction includes fighter (and some missile) air defense, some garrison troops, some elements of the surface navy, and not much else. The chief physical threat to the people and territory of the United States, enemy nuclear missiles, finds us *without any direct defense at all.* Do not think that one is talking here about a *perfect* defense being lacking! Many people assume that in nuclear war as in all others there is a kind of standoff, with a small margin of error in which the few weapons which hit from one side or another, might swing the tight balance. Not so; there is no missile defense at all. Most of those nuclear warheads which are in fact launched against the United States will hit their targets. Our defense is wholly based on an

---

of our launch systems, at ample scale. (The refuelling tankers of the long-range bomber force are in part kept, but placed in the service of the general-purpose forces.)

ii) Combat aircraft are kept in full strength, assigned to the defenses of the U.S., Europe, and Japan.

iii) The ground forces become aimed more clearly at defense of Europe. The Marine and Navy power projection forces, the carriers and the amphibious landing groups, are sharply cut.

iv) The Navy surface ships, freed from protection of carriers and landings, still provide an ample bluewater force of more modern ships second to none. Navy and Marine combat air, converted mainly to basing on land, are now to be devoted to the air defense of Europe and of Japan.

v) The anti-sub submarines are numerous enough for sealane defense, but not so many as to threaten the Soviet undersea deterrent.

vi) Airlift remains as it is, in support of NATO. Proposed deployments are shown in Figures 50, 51, and 53.

DEFENSE SECRETARY FY 1978.

indirect effect; the threat of swift and terrible retaliation (or, in fact, the possibility of the United States striking first with nuclear weapons against a war measure of another kind begun by others).

In the nuclear case, the U.S. capability for actual defense is no different from that of other powers. Nobody can be said to have any defense against nuclear war save the indirect sort, by retaliation, by being allied, and so on. In nonnuclear war, the case is quite different. The long borders of other states often find antagonism on the other side, for example in Central Europe or along the Amur River or the plains of the Indus. Not so for the United States. Our possible enemies lie across great oceans. We therefore need not mount powerful border forces and heavy inner garrisons, hardly against Canada, Mexico, Cuba, or the Bahamas! Yet we have conventional military power by land, sea, and air, which is inferior to none. Those forces are intended to engage others far away from U.S. shores, in defense of the United States only indirectly, by the defense of our allies, especially NATO and Japan.

This strongly affects the physical nature of U.S. forces. The U.S. Navy is alone among navies in the ability to fight in powerful air combat almost anywhere without land bases, using giant aircraft carriers. U.S. troops are provided with a very powerful airlift capability, for ocean-spanning no less than motion about any theater of war. The Navy and the Marines possess amphibious forces capable of landing on shore lines, whole oceans away from the bases at Norfolk or San Diego. The United States is really the only world military power in that sense; the U.S.S.R. is still a distant second.

The facts and figures behind all this, the history of its evolution, and a critique of its use, past and potential, are explained at length in our book. Here we simply remark that this sense of military projection, of readiness for combat anywhere, of unique preparations and plans, can be described only partially as the prudent preparation of a nation anxious to spare its own land and people the terror of war. Of course that is true, but it is equally plain that such a power appears to others in a very different light, namely, a light whose glow comes in part from the flames of Vietnam, Cambodia, the Bay of Pigs. . . .

Is a forward stance the best and safest defense? Is the preparation of a full-scale, mobilized, ever-ready, active force always the best way to defend one's interests in peace and wisdom? This book tries to put the case that those easy axioms are wrong. Even football teams recognize that a strong offense is not the only part of the

MANPOWER IN THE PROPOSED FORCES

	1978	Proposed
US Army	790,000	550,000
US Navy	540,000	350,000
US Marine Corps	190,000	75,000
US Air Force	570,000	450,000
TOTAL	2,100,000	1,400,000

FIG. 80

What the manpower of the active armed forces of the United States would be under our proposals, some time in the 1980s. The Marine strength, now held mainly for amphibious and interventionary uses, is sharply reduced. The Army is cut, to the same end, but retains most of the heavy forces aimed at defense of Europe. The Air Force retains full tactical strength, losing that rather small number of persons who serve the vulnerable airborne and land-based nuclear weapons. The Navy retains strong strategic and sea control forces, is much reduced by near elimination of the expensive and vulnerable power-projection forces, though the combat air-wings off the carriers are retained. The strategic defense and warning systems are kept, as well as most intelligence capabilities.

DEFENSE SECRETARY FY 1978 and MANPOWER FY 1978.

game; war is more complex still, and war today is in fact itself risked by the very preparations which we are told are made in the interest of defense. The steady rise in the power of war machines and the equally steady increase in the anxiety of the world over the full-scale war which all have until now avoided are intimately connected. So are the innumerable smaller but locally devastating wars the world has not avoided. The "worst case" assumptions and the "prudent" increase in power can become themselves dangerous; we argue that they, in fact, have.

The policy we propose is not based on any simple axiom of military preparedness. It is based on a serious issue-by-issue and weapon-by-weapon examination of the world of warfare of today and the near future. We reach the conclusion that a very different structure of military force would better meet the ends which Americans seek, and we seek to describe how that structure might be. (Consult Figures 79 and 80.)

WHAT THE CHANGES WE RECOMMEND WOULD DO

We believe the forces we outline in the book would carry out with reliability the following aims, which we see as attracting a large consensus of American opinion, within the bounds of what is fully feasible.

The new forces would defend directly and indirectly, as at present, all U.S. territory and the territory of U.S. chief allies; that is, the NATO powers and Japan, against any plausible dangers of the next decade.

The new forces are safer forces, for they would not impel in others, who may misread American purposes, so great a self-preparation. They do not retain the forward projection, the "over-kill," the swift worldwide intervention abilities of the present forces. It is long past time to try this needed gain in safety; the uniform increase in means and in tensions that the past generation has watched has continued too long.

In a word, the new forces represent a decrease in the militarization of the United States and of the world, because the U.S. contributes such a major, dominating share to world militarization.

The new force limits the capacity of the United States to intervene in remote parts of the world, where interests not vital but nevertheless influential have used national forces often for private ends.

The new forces are fully effective, though they represent a decreased threat of mortal injury and enduring harm to civil populations and to indirectly involved states.

The new forces would be visible and understandable, to a degree more susceptible to eventual agreements of international arms control.

The new forces would release powerful economic energies in our own country, because they demand so much less support. This can be seen as a large dollar saving to the federal budget—$50 billion or more yearly—or as a way to reorganize to provide more jobs, by directing effort to less capital-intensive activities than the remarkably complex technologies of modern war.

The new forces would serve as an example to others, a morally and more nearly defensible stance for our country, a sign of hope for a world not under the shadow of destruction. World response is beyond our control, but the United States can with safety take the first steps, the steps which the leader and the driver of the incessant race for more and better arms—our great country—should make.

WHAT THE CHANGES WE RECOMMEND WOULD NOT DO

We do not believe that these more reasonable forces would in any way endanger the security of the United States or of any of its allies. We have provided for the effective fulfillment of all American pledged commitments. We submit that under those circumstances no trace of the sense of isolationism adheres to the scheme; we are not alone as a nation in this world, and we do not intend to "go it alone." The force level we propose would much reduce the ability of the U.S.A. to extend its military force by intervention to places where we do not now have such preparations. But of course it would not eliminate that power. This country has great economic and political weight to use everywhere. The forces in our new proposal are still large in order, reasonably, to fulfill U.S. commitments to Europe and Japan—they are so large that a small fraction of them is a major force in most of the world.

While we believe Uncle Sam would be a more moderate fellow, not so able to "tilt" by threats in the Indian Ocean today or off the coast of Chile tomorrow, he would be by no means tied down. To those who see in the history of U.S. foreign interventions an unworthy imperialism, we can offer the surety that surplus military forces and preemptive or immediate response would become more costly, more open to public debate, more needing consensus. But if Uncle Sam wants to shoulder a stick, he would still retain a big one.

Only political decisions can make so powerful a society as ours an unquestionable good neighbor. On the other hand, for those who hold that we must be able to act by widespread and major intervention, we reply that public debate and Congressional action can indeed supply large forces for such service. They will, however, not be at the quick disposal of decisions made in private. Military establishments are not the only, nor indeed, the chief keys to national action. They are only its instruments; yet too big a stick is *not* defensible, and it can increase the chance of fights, not least by its effects on others.

A public watchword can be "Remember Pearl Harbor." What Americans mean by that is clear; but from the point of view of its Japanese military organizers, it was a brilliant success, audacious, technically advanced, working great damage on the enemy at small cost. The attackers paid later; the biggest ships they sunk were in the end of a type never useful throughout the war, and the move guaranteed the defeat of the Japanese Imperial Forces by making certain that the United States came into the war at once. The very treachery and success of the attack unified the determination of the previously uncertain American public. Pearl Harbor was, on balance, not success but ruin for the Japanese Imperial Forces. One can sense the military cast of thought which made Pearl Harbor— we do not include the planned illegalities—in the overestimates and overconcerns of many of our own planners. They prepare against the shock of the unprepared defense of Pearl Harbor—let it not happen again—but they tend to forget its greater strategic lesson.

THE PROCESS OF TIME

The arms race of the postwar world is now more than twenty-five years old. It does not seem to reach a clear limit, whether in the quantity of arms, in their spread through the world, or in their nature. Yet if the race does not slow down, it is hard to see the world remaining for a long time free of a general and unprecedented catastrophe. To this problem we offer a straightforward solution in detail. The unquestioned leader of that race—the United States—is in the best position to slow down. Let the U.S.A. begin to reduce its forces as far as it can *without* real risk to its security. That cannot be done overnight. But it can be done after debate, legislation, change in the climate of opinion, done step-by-step, in this base or in that, in this dockyard or that airfield, in this withheld production or that deferred research.

Obviously equity, safety, and common sense require a sure and measured pace in this reduction, in this calling a halt to the race for arms. We have given here and there particular reasons for a period of years in each action, a rate of reduction, or a time after which we would remove some system. We do not summarize all these here, but we can say that our study is aimed at an indefinite period between now and the end of the eighties. The dates are less important than the double goal: something real and tangible soon, to give an earnest indication that the scheme is real; and equally important, that the greatest military organism in the world cannot and ought not in equity diminish overnight.

WHAT NEXT?

It is clear that these proposals for action by the United States, not in the first instance dependent on what others may or may not do, are not the final word. They are a set of *first steps* in a new future for international affairs in this world of nuclear weapons and constrained growth. They will of course have to be complemented by a diplomatic policy and an economic attitude toward the world which will take them into account. This volume is in no position to describe or outline what those policies may be. They will of course depend a great deal on the reaction of other nations, the most important of which are those nations against which our forces are so strongly polarized. Those countries are not transparent for analysis, even were the authors competent to attempt the task.

Let us go ahead, prudently, seriously, openly, even generously, in the knowledge that the struggle against an automatic and growing militarism is a problem for most of humanity. Evidently, NATO lands and Japan are directly concerned; but all countries will pay attention. The pitfalls lie on both sides of the path: If we in the United States do not act for ourselves, it is hard for others to take the initiative, since we usually call the tune. Multilateral procedures may come into being once the thaw begins, but up to now only a little more than atmospherics has been gained from the tedious diplomacy of arms control. Our proposal opens another road, a road the nations need; we see no other open way.

The minute is late. The hands of the nations' clock may inch forward and back year by year, but over the years they have crept toward midnight. That clock of war can be slowed. Now is the time to wind down!

TO LEARN, TO DECIDE, TO ACT

"The function of information is to make decisions."
John R. Platt, 1956

Our book is about what should be done in redirecting the military affairs and foreign military policies of the United States, with attention to the immediate consequences of the changes we recommend. But producing a book does not in itself change policy—to change policy, individuals must become convinced, and then must begin acting on their convictions within the scope of their life's position. This epilogue outlines these processes of change.

First of all, to read a book is not necessarily to become convinced; one may read for entertainment, as one reads the daily newspapers, without the slightest intention of taking part in the events described. But to act, one must read with judgments pro or con, and determine whether one agrees enough to support the conclusions actively. Appropriate actions can then be discussed for three major readerships: the professionals, the politicians, and the public.

Professionals within the Department of Defense outnumber disarmament professionals by thousands to one. Their influence is so dominant that their support is of paramount importance; officers and career officials within the Defense Department should take those military policies advocated here with which they can agree as the basis for official recommendations to the chain of command. We have proposed nothing that could not reasonably be put forward by the Chiefs of Staff and the Secretary of Defense; such advocacy, impelled by the recommendations of professional colleagues, would greatly facilitate the changes suggested here. Disarmament professionals in the academic world and the U.S. Arms Control and Disarmament Agency can for their part focus their efforts on contributing to comprehensive alternatives to present policies.

Politicians concerned with American military policy work in the White House, the Senate, and the House of Representatives. At the apex, the President as Chief Executive and Commander-in-Chief and the Secretary of Defense as his principal adviser for military affairs could exert a preponderant influence if convinced that the changes recommended here are authentic and should be followed.

More numerous, more diverse in opinion, and often more open to influence, Senators and Representatives can use their power of the purse and their authority to make rules for the regulation of the land and naval forces to help put our recommendations into effect. The influence of members of Congress is expressed at different times of the year in eight major committees, as well as in the two Congressional bodies meeting as wholes. This process is broadly outlined here.

HIGHLIGHTS OF CONGRESSIONAL ACTIVITY PRECEDING A NEW FISCAL YEAR

January:
Budget Committees begin work on total budget and major categorical limits (i.e., national defense)
Armed Services Committees begin hearings, etc. President submits his annual budget.

by March 15:
Armed Services Committees submit spending recommendations to Budget Committees.

by April 15:
Budget Committees report recommendations (First Concurrent Resolution) to each House.

by May 15:
Armed Services Committees have reported all authorization bills to House and Senate.

Congress (House and Senate agreeing) votes on First Concurrent Resolution, including a total budget and categorical limits.

May 15 to September:
House and Senate Armed Services Committees make final recommendations for military budget.

Congress votes authorizations for military.

President signs or vetoes.

House and Senate Appropriations Committees (and especially their Defense Subcommittees) recommend appropriations for the military.

House and Senate pass appropriations for military; differences are reconciled.

President signs or vetoes.

by September 15–25:
House and Senate adjust appropriations and total budget limits to be mutually compatible.

President signs or vetoes.

October 1:
New Federal fiscal year begins.

The process starts each year in the new Budget Committees of the Senate and the House, where budgetary priorities and limits are placed on all major appropriation categories, including the military. Simultaneously, the Armed Services Committees of House and Senate meet to determine the specific budget for the military. The key significance of these two committees has drawn a membership both of advocates and critics of increasing military expenditures. Minority reports from either of these committees would have great influence in rallying support both within and outside the Congress for changed military budgets.

Somewhat later in time the full House and Senate debate, amend, and vote authorizations (maximum amounts) for the programs and expenditures connected with the military. Though this is usually the decisive stage in approving budgets, these authorizations must still pass through the Appropriations Committees of House and Senate where the total Federal budget is scrutinized, and then back to the full House and Senate for the final votes actually appropriating the money, never more than the authorizations and sometimes less.

Somewhat on the sidelines sit the House International Affairs Committee and the Senate Foreign Relations Committee. Though large military budgets are rationalized by the supposed foreign policy goals of the United States, these committees have as yet not found an opening for the systematic critique of the executive branch's expensive conduct of international affairs. These committees remain potential opportunities to examine America's foreign policy and its overseas military forces.

Out of this maze of legislative detail one overriding principle should be emphasized—there are many, many opportunities in Congress to examine, criticize, and change American military policy and practice.

Public debate and action in a democracy are the final test for major government action. Unfortunately, the mystique of the secrecy which surrounds military action in wartime has extended to inhibit public examination of military expenditures in peacetime (see, however, the Note on Secrecy). Furthermore, while specialized groups lobby for appropriations as vigorously on the civilian side as on the military, only once in a while does the American public as a whole demand some major change in the normal direction of government. The New Deal and the end of the war in Vietnam represent change on the scale appropriate to changing the military practice of the past 30 years.

How can such a large-scale change begin? The change must begin with an individual questioning the massive amount of spending on our military rather than abdicating political responsibility to "the experts." The change must continue with an individual's act and with an act of faith. The individual's act must be to communicate to Representatives and Senators in Congress, whose task it must become to reduce the military budget in ways specified, for example, in this book. The act of faith is simply the belief that others will do the same.

A telephone call to a Representative is surprisingly inexpensive and effective; a letter or Mailgram is also excellent.* Inquire whether your Representative has read this book, and what he or she proposes to do about its recommendations. Your two Senators can be approached in the same way. Their staff will be encouraged or directed to take serious thought on these military matters by your request, and you will receive a courteous if sometimes cautious reply. Next you can ask some close friends to learn and spread the message—whether through this book, by pamphlets, or by word of mouth—that military expenditures must be reduced and the military emphasis in the conduct of American foreign policy ended.

Beyond the individual's act and the first contact with friends come a series of public acts that merge with politics. We list only a sampling here: the systematic canvass of all one's acquaintances; the locating and organizing of those particularly willing to help into some permanent grouping for action; helping a particularly responsive Representative or Senator win reelection; actively searching for and coordinating with other groups doing the same work†; seeking to convince other organizations of the overriding importance of reducing the military component to achieve their own goals for society.

---

*The address: The House of Representatives, Washington, D.C. 20515. The Senate, Washington, D.C. 20510

The telephone for all Senators and Representatives: (202) 224–3121.

†One such group is the Coalition for a New Foreign and Military Policy (120 Maryland Avenue, N.E., Washington, D.C. 20002) which represents a wide range of church groups, unions and social action agencies in the effort to transfer major federal emphasis from military foreign policy to domestic, civilian needs. They maintain a Washington base, essential for any effort to change national priorities. They furnish pamphlets, voting records, and the schedule of votes in Congress, which are necessary to pinpoint citizen effort for greatest effect.

Wars cannot end war; we cannot fight for peace. In this country we can communicate with our government, and we can communicate with each other. "Let us raise a standard to which the wise and honest can repair"—that is the tribute a democracy asks from its own people.

SOURCES AND NOTES
Annotation and Short Titles of Main Sources

Among the books and documents described below, those which are cited frequently in the chapter notes following are referred to by a shortened title as indicated. For annual publications, the year which appears in the proper title is included in the short title.

A. OFFICIAL DOCUMENTS

Most of the information we have relied upon comes from the voluminous and detailed publications of the United States Federal Government. Defense policy is of course made by the Executive branch of the government, mainly the Department of Defense. The Legislative branch—that is, the Congress—also plays a role. This is done primarily through the "power of the purse," the process by which Congress allocates the funds needed to carry out the Defense Department's plans (to pay and house the required numbers of military personnel, manufacture planned numbers of weapons, develop new equipment, conduct exercises, and so on).

The annual Congressional review of the defense budget is the single most important source of information on U.S. defense policy; this is because the most authoritative and comprehensive public source materials about U.S. military forces and spending are those prepared annually by the Defense Department for its budget presentation to Congress.

The Defense Department budget is part of the overall Federal Government budget (see U.S. GOVT BUDGET below). The President's Office of Management and Budget prepares the budget document during the summer and autumn of each year, working together with the various Federal departments and agencies to modify and coordinate their requests. The budget (in several volumes) is then submitted to Congress in January and February. Over the next six months or so, the budget is divided into its departmental components for review by specialized Congressional committees, in public "hearings" of departmental and other witnesses. The committees then "report out" their parts of the budget—with amendments, deletions, and additions—to the two houses of Congress, which finally vote to pass, amend, or reject the committees' versions.

The accounting year which the budget covers is called the fiscal year (FY). Through 1975–76, the U.S. Federal Government's fiscal year began on 1 July and ended on 30 June. Fiscal year 1975, for example, began on 1 July 1974 and ended on 30 June 1975. In 1976, the Federal accounting period was changed (for the first time in U.S. history) to 1 October–30 September. After a transition in the summer of 1976, fiscal year 1977, the first under the new accounting period, covered 1 October 1976–30 September 1977.

Most government documents are available from the Government Printing Office (GPO) in Washington, D.C. (zip code 20402). They are also stored in depository libraries throughout the country. (It helps to seek an expert librarian's advice when using archival collections.) Some of the Defense Department documents listed below are too specialized or informal to be printed and stocked by the GPO. They may be located in the libraries of the Department of Defense, the Department of State, the U.S. Arms Control and Disarmament Agency, or the Library of Congress. These documents and other information are also generally supplied by the Defense Department on request and without charge to citizens who show serious interest. One can write to the helpful Office of the Assistant Secretary of Defense for Public Affairs, Department of Defense, Washington, D.C. 20301. The telephone number there is 202–697–5737.

(1) *Department of Defense*

The *Annual Defense Department Report* for each fiscal year is presented to Congress by the Secretary of Defense. Sometimes called the "Posture Statement," this volume seeks to explain and to justify the coming budget request, and is a detailed survey of the forces and support systems as they are and as they are planned to change. Its detailed emphasis varies yearly as changes in interest and scale occur for the many parts of the structure, but it is always comprehensive. From FY 1949 through FY 1968, the Secretary of Defense made two yearly reports: an annual budget statement to Congress, prior to the coming fiscal year; and a retrospective historical report on the major events of each fiscal year, published one or two years after the fact. The historical report, rather than the budget statement, was called the *Annual Report;* it satisfied a legal requirement made by Congress for such a report. Since FY 1969, material from the two types of report have been combined in one document.

–abbreviated: DEFENSE SECRETARY FY 19—

The *United States Military Posture* is a brief annual report by the Chairman of the Joint Chiefs of Staff. It emphasizes the forces of the U.S.A. in comparison to those of the Soviet Union. Less attention is paid to support structures, and its interest is not in budgets so much as in purposes and uses for weapons and manpower. The term "Posture Statement" is often confusingly used for this document as well. Thus, we have avoided the term in both shortened titles.

–abbreviated: JOINT CHIEFS FY 19—

The two citations above are only the most general of the resources available from the Department of Defense. We have used also the more specialized reports, such as those dealing with manpower, research and development, weapons systems costs, and other support matters. In addition there are annual reports of the Secretaries of the individual services. Here we will cite and give abbreviations only for those specialized publications to which we have had occasion to refer several times. The others we cite in full in the notes where they apply.

Manpower Requirements Report, issued annually and prepared by the Office of the Assistant Secretary of Defense for Manpower and Reserve Affairs.

–abbreviated: MANPOWER FY 19—

Program Acquisition Costs by Weapon System, issued annually and presenting cost estimates of major weapons acquisitions.

–abbreviated: PROGRAM ACQUISITION FY 19—

Program of Research, Development, Test and Evaluation, issued annually as the Congressional statement of the Director of Defense Research and Engineering.

–abbreviated: PROGRAM RESEARCH FY 19—

(2) *Congress of the United States*

Primary documents, including the text of many of the full Defense Department reports, are the hearings conducted each year in both houses of the Congress by the Armed Services Committees and by the Appropriations Committees. These are bulky but rich. Often the interrogation brings out valuable material. We cite them throughout the notes where most relevant.

Congressional Budget Office (CBO). This staff research arm of the Congress has in the last couple of years become a most valuable source of information and analysis of all Federal policies. Of all sources we know, their recent series of ''Budget Issue Papers'' come closest to the intent of our book, combining introductory technical material, specific data, and policy analysis in one coherent text. The papers differ from this book in two ways, of course: Each is centered on a single issue, and the CBO papers do not end with any clear advocacy of one or another conclusion. Rather they outline various positions which span the alternatives. We cite a few studies we have found useful here; there are many more and they continue to be issued periodically. The CBO ''Budget Issue Papers'' are all published by the U.S. Government Printing Office.

U.S. Strategic Nuclear Forces: Deterrence Policies and Procurement Issues, April 1977.

Planning U.S. General Purpose Forces: Overview, January 1977.

Planning U.S. General Purpose Forces: The Tactical Air Forces, January 1977.

The Costs of Defense Manpower: Issues for 1977, January 1977.

The Military Retirement System: Options for Change, January 1978.

Counterforce Issues for the U.S. Strategic Nuclear Forces, January 1978.
> –abbreviated: CBO: (short title, month, year)

Library of Congress. The research arm of the Library is the Congressional Research Service, another source of special studies of continuing interest. The studies are done on request of Congressional offices.
> –abbreviated: CRS: (short title, month, year)

(3) *U.S. Arms Control and Disarmament Agency* (ACDA)

This small Executive agency produces a number of relevant reports and booklets, maintains a mailing list for interested citizens, and will answer public inquiries. One work found helpful is *World Military Expenditures and Arms Trade,* issued annually.
> –abbreviated: ACDA: (short title, month, year)

(4) *U.S. Department of State*

The foreign policy arm of the U.S. Government, the Department of State produces a voluminous amount of documentation, speeches, and reports, most available to the interested citizen.
> –abbreviated: STATE: (short title, month, year)

(5) *General statistical material for the United States*

The *Bureau of the Census.* Produces the two volumes, *Historical Statistics of the United States: Colonial Times to 1970,* GPO, 1975.
> –abbreviated: HISTORICAL STAT OF THE U.S.

The *Office of Management and Budget.* Formerly the Bureau of the Budget, this produces the *Budget of the United States Government* annually.
> –abbreviated: U.S. GOVT BUDGET FY 19—

B. MILITARY HARDWARE

There has grown up a substantial and continued encyclopedic literature describing, with varying degrees of reliability and precision, all the world's military weapons from hand grenades to thermonuclear missiles and aircraft carriers. These are indispensable though they require a somewhat skeptical sort of use.

The most complete and compendious form a series of yearbooks: *Jane's All the World's Aircraft, Jane's Fighting Ships,* and *Jane's Weapons Systems* (includes guns, missiles, and armored vehicles). These large expensive volumes are found in good reference libraries; they are issued every year and include specifications, drawings, history, and dispositions. The books originate in London but are published in the United States by McGraw-Hill Book Company, New York.
> –abbreviated: JANE'S AIRCRAFT 19—
> JANE'S SHIPS 19—
> JANE'S WEAPON SYSTEMS 19—

Many readers will want to use the several little paperbound abridgments of the big Jane's volumes, much smaller and cheaper, and still very useful. These are called *Jane's Pocket Books of . . . Major Combat Aircraft . . . Missiles,*

. . . *Modern Tanks and Armored Fighting Vehicles,* et cetera. Their publisher is MacMillan Publishing Company, New York.

Two very useful little books review the most important aircraft in some detail: *Bombers in Service, Patrol and Transport Aircraft since 1960;* and *Fighters in Service, Attack and Training Aircraft since 1960.* Both are by Kenneth Munson, Macmillan Publishing Company, New York. They appeared in revised American edition in 1975.

C. MILITARY POWER SURVEYS

Two international organizations produce annual compendia of importance, covering the world military situation. The *International Institute for Strategic Studies* (IISS) is a private institute in London publishing both special studies and annual reference works. One of its annual works is a tabular account of the deployed and reserve military forces—all branches, with their major equipment—for all nations, from Afghanistan to Zambia. Various tables of equipment, comparisons of alliances, and the like round out a very valuable paperbound book (available for about $5 from the Institute, 18 Adam Street, London, WC2N 6AL). The title is *The Military Balance*.

–abbreviated: MILITARY BALANCE 19—

The IISS also publishes *Strategic Survey* (annually), *Survival* (bimonthly), and *Adelphi Papers* (periodically).

The *Stockholm International Peace Research Institute* (SIPRI) also publishes both annual and occasional studies on problems of peace and conflict. It is a similar research institute, but is financed by the Swedish Parliament and is internationally staffed and governed. The most useful annual is its *Yearbook* which reviews the year before the date of issue with respect to the arms race worldwide. It usually contains rather detailed studies on some issues exemplified by the year's events, plus careful tabulations of world expenditures, arms trade, arms limitation treaties, nuclear tests, and the like. *World Armaments and Disarmament: SIPRI Yearbook* is distributed in the United States by Crane, Russak and Co., 347 Madison Ave., New York, N.Y. 10017.

–abbreviated: SIPRI YEARBOOK 19—

In late 1975 there was published an integrated collection of a number of individual sections from various SIPRI Yearbooks, appearing as a single book of modest length. It is a very useful introduction, entitled *Arms Uncontrolled*. Prepared by Frank Barnaby and Ronald Huisken, it is available from Harvard University Press, Cambridge, Mass.

–abbreviated: ARMS UNCONTROLLED

The *Brookings Institution* in Washington D.C. (1775 Massachusetts Ave., N.W., 20036) is a well-known private research institute. It does work in economics, government, and the related social and policy sciences and publishes one continuing series of brief special studies which we have found very useful. The series is called *Brookings Studies in Defense Policy.* We list a few of them, cited often in our book.

Martin Binkin, *Support Costs in the Defense Budget*, 1972.

Martin Binkin, *U.S. Reserve Forces: The Problem of the Weekend Warrior,* 1974.

W.D. White, *U.S. Tactical Air Power: Missions, Forces, nd Costs,* 1974.

–abbreviated: BROOKINGS: (short title)

D. THE LITERATURE AS A WHOLE

There is a large literature, both serious and popular, in this entire area. We cite the books and periodicals individually as we use them throughout the chapter notes. Particularly useful periodicals include:

Aviation Week and Space Technology, Washington.

Bulletin of the Atomic Scientists, Chicago.

Foreign Affairs, New York.

Foreign Policy, Washington.

International Security, Cambridge.

Scientific American, New York.

Survival, London.

The newspapers of main use here are the *New York Times* and the *Washington Post.*

Two special periodicals of relevance are:

Arms Control Association, *Arms Control Today,* Washington.

Center for Defense Information, *The Defense Monitor,* Washington.

One book which takes a view of the whole matter for the general reader, with much data and analysis, but with not much effort to arrive at judgments differing from the *status quo,* is the work by a professor at King's College, London, Laurence Martin, *Arms and Strategy: The World Power Structure Today,* New York: David McKay Company, 1973. It is also well-illustrated.

—abbreviated: ARMS AND STRATEGY

An excellent comparative analysis of worldwide military and social and other statistics is provided by Washington economist Ruth Leger Sivard in *World Military and Social Expenditures* (published in volumes of 1974, 1976, 1977, and 1978), Leesburg, Virginia 22075: WMSE Publications, P.O. Box 1003.

–abbreviated: WORLD MILITARY AND SOCIAL EXPENDITURES 19—

TEXT SOURCES AND NOTES

We hope that we have put these notes in a form useful for the general reader. We have included not only sources necessary and helpful for additional research but also asides not indulged in within the main text. In each note we begin with a phrase which orients the reader toward the topic under discussion; we have thereby allowed the main text to read freely without footnote markers of any kind. Consulting the notes for the chapter and page will lead you at once to a source or comment, without much back reference.

CHAPTER 1

page 3

The emergence of cold war political views and military forces under Truman and Eisenhower is discussed in Samuel Huntington's *The Common Defense* (New York: Columbia University Press, 1961), an engrossing account of U.S. military policy from 1945 to 1960.

Since the early 1970s, there has been an unusual opportunity to reconsider the heavily mobilized peacetime stance adopted by the USA after the Korean War and not much changed since. From the mid-1950s until 1973, public debate was preoccupied with three inflammatory issues: nuclear testing in the atmosphere, ABM (anti-ballistic missile) deployment, and the Vietnam War. Although these issues had far-reaching implications, they were directly concerned with only a limited part of U.S. military policy. Surprisingly, the current political breathing space, an opportunity to take a larger and longer-term view, has been recognized and exploited not by those who favor arms reductions, but by those who support the status quo.

What is probably the earliest notice of this special time is given in DEFENSE SECRETARY FY 1975, prepared in late 1973 and published in March, 1974. In a section of that document entitled "Settling Down for the Long Haul," former Defense Secretary James Schlesinger states (p. 15):

> As I mentioned at the outset of this introduction, this is our first peacetime defense budget in a decade. It is, therefore, an appropriate time to consider how best to settle down for the long haul, for the

continuing, steady task of providing an adequate defense for the United States and its interests. During the next few years, we must search for and assess the best R&D, weapons acquisition, and other strategies for the long haul. In doing so, we must ask such questions as how we can most efficiently compete with our major potential opponents, and what constitutes our own major strengths and weaknesses.

page 4

In political science, there is a body of literature concerning "bureaucratic politics." This covers, in part, the tendency of large bureaucracies, particularly Federal Government departments and most especially the very large defense bureaucracy, to work for self-aggrandizing or self-perpetuating policies, even when these policies diverge from the public interest. A good general treatment of such not uncommon bureaucratic inertia is A. Wildavsky's *The Politics of the Budgetary Process* (Boston: Little, Brown and Company, 2nd edition, 1974). Usually this literature does not attempt to prove the most obvious thesis—that, in comparison with the population at large, military employees (career military officers more than civilian officials in short-lived appointive positions) tend to support larger military budgets and greater reliance on military force as a tool of policy. This "turf-protecting" behavior of defense bureaucrats is, however, illustrated in studies of inter-service rivalry (among the Army, Navy, and Air Force) and inter-bureaucratic competition (among the Defense Department, State Department, National Security Council, Office of Management and Budget, and the Arms Control and Disarmament Agency, among others). Two noteworthy studies of inter-service rivalry and its impact on the overall defense budget are: P. Hammond, "The B-36-Carrier Controversy," in W. Schilling, P. Hammond, and G. Snyder, *Strategy, Politics and Defense Budgets* (New York: Columbia University Press, 1962); and "Missile-gap Mania," in H. York, *Race to Oblivion* (New York: Simon and Schuster, 1970). M. Halperin describes inter-bureaucratic competition in *Bureaucratic Politics and Foreign Policy* (Washington: Brookings Institution, 1974). In addition, several studies have been done on the influence of local defense industry spending and employment considerations (as opposed of foreign and defense policy concerns) on Congressional votes for increases or decreases in the defense budget. These include: "Defense Spending and Senatorial Behavior," in B. Russett, *What Price Vigilance?* (New Haven: Yale University Press, 1970); A. Kanter, "Congress and the Defense Budget: 1960–1970," *American Political Science Review* 66 (March 1972), pp. 129–142; and C. Goss, "Military Committee Membership and Defense-Related Benefits in the House of Representatives," *Western Political Quarterly* 25 (June 1972).

page 5

Exaggerated claims about a Soviet military threat are analyzed in Chapter 4.

page 5

The seven million constituents on the defense budget payroll include (in round numbers): two million uniformed military personnel; one million civilian employees of the Defense Department; one million reservists; one million retired persons; and two million employees in the defense industries (companies in the private sector which manufacture weapons and other military equipment). More detailed estimates of these groups are given in the annual Defense Department publications, MANPOWER report and *Selected Manpower Statistics* (Directorate for Management Information Operations and Control).

page 7

The dangers of nuclear weapons and the lack of defenses against long-range ballistic missiles are discussed in Chapter 2 and in Part II of this book.

page 7

The military budgets of West Germany, France, and other countries are given in the annual publications, SIPRI YEARBOOK, WORLD MILITARY AND SOCIAL EXPENDITURES, and MILITARY BALANCE. The two latter surveys also give numbers of men in the armed forces. MILITARY BALANCE gives information on land-army size and equipment as well.

page 9

The counter-productive aspects of international arms control negotiations are discussed in G. Rathjens, A. Chayes, and J. Ruina, *Nuclear Arms Control Agreements: Process and Impact* (Washington, D.C.: Carnegie Endowment for International Peace, 1974).

page 11

Long-term changes in the quantity of resources allocated to the military are described in SIPRI YEARBOOK 1968/69.

CHAPTER 2

page 12

The study of military capabilities involves looking at both qualitative factors (measures of the performance and vulnerabilities of particular weapon systems or combat units) and the quantities of weapons of various types. The most useful source of information on the former are the various JANE'S reference works; on the latter the MILITARY BALANCE.

The single most useful source of information on U.S. military forces and their rationales is the annual DEFENSE SECRETARY report.

page 15

The inherent conflict between the primary and secondary functions of U.S. strategic nuclear weapons, debated in the 1950s, has received attention recently in an as yet unpublished study by Earl Ravenal of the Institute for Policy Studies in Washington. Ravenal argues that the United States should forego any attempt to maintain extended deterrence as part of its nuclear weapons policy; and he claims that, given the danger it presents to the U.S. population, it has never been a truly believable ("credible") aspect of U.S. policy throughout the post-war period.

U.S. commitments under the various arms control treaties negotiated during the twentieth century are described in ACDA, *Arms Control and Disarmament Agreements*.

page 16

The Coast Guard fleet is described in S. Morison and J. Rowe, *Ships and Aircraft of the U.S. Fleet* (Annapolis, Maryland: Naval Institute Press, 10th edition, 1975).

The best short description of the major components of the U.S. general-purpose forces is given in the annual MILITARY BALANCE.

page 18

U.S. foreign military deployments are authoritatively outlined in two Defense Department publications: the annual *Selected Manpower Statistics* and the *Base Structure Annex* to the MANPOWER FY 1978.

page 21

The last part of Chapter 2 summarizes arguments which are documented and developed at much greater length in Part III of the book, to which the reader is referred for evidence and sources.

CHAPTER 3

page 23

This chapter analyses and synthesizes quantitative information on the world's military forces from two main kinds of sources. The first is studies done by R. Forsberg over a period of several years at the Stockholm International Peace Research Institute on the topics of world military spending, world military research and development resources and projects, and U.S. and Soviet strategic nuclear weapons developments over the last 15 years. The second is information presented in the last several editions of the annual MILITARY BALANCE and WORLD MILITARY AND SOCIAL EXPENDITURES.

CHAPTER 4

page 35

Comprehensive but concise background material on the Soviet Union, for the interested reader, may be found in *Political Handbook of the World,* published annually (New York: McGraw-Hill) and in the U.S. State Department's *Area Handbook for the Soviet Union* (Washington, D.C.: U.S. GPO, 1971). Of course, a voluminous literature exists on the subject.

page 37

For a recent overview of Soviet-American relations and U.S. policy, see "Overview of U.S.-Soviet Relations," testimony before Congress by Marshall D. Shulman, Special Adviser to the Secretary of State on Soviet Affairs (STATE, Bureau of Public Affairs, October 26, 1977).

More detailed outlines of Soviet forces are given in the annual MILITARY BALANCE.

For interpretations of a heightened Soviet threat in the 1970s, the reader is referred to trade journals such as *Aviation Week and Space Technology,* to Congressional testimony on Soviet-American relations and on the Strategic Arms Limitation Talks, and to pamphlets of lobby groups such as the Committee on the Present Danger.

Some alarmist allegations of the 1970s have compared the Soviet military programs to prior wartime buildups. A detailed report by Representative Les Aspin alleges this to be unfounded and states that Soviet military expansion in recent years "has been far less dramatic than many scare stories would lead us to believe." See "What Are the Russians Up To? A perspective on Soviet military intentions," November 28, 1977.

page 40

The JOINT CHIEFS report, issued annually in January, deals mostly with the Soviet-American balance.

The quote of the Joint Chiefs of Staff regarding nuclear superiority is taken from a January 28, 1977 letter of General George S. Brown, Chairman of the Joint Chiefs, to Senator William Proxmire; it is reprinted in *Survival,* v. 19, n. 2 (March–April 1977), pp. 76–78.

Further evidence on the continued stability of the nuclear forces may be found in the annual Congressional report, *Allocation of Resources in the Soviet Union and China* (Joint Economic Committee). In the 1977 report, government officials testified that the newest generation of Soviet intercontinental ballistic missiles were still less accurate than the present U.S. systems. See also U.S. Congress, Senate, Committee on Foreign Relations, *United States/Soviet Strategic Options* (Washington, D.C.: GPO, March 1977). In March, 1977, Secretary of Defense Harold Brown stated that "with respect to the strategic balance I believe that we and the Soviets are in a situation of rough equivalence . . . We are ahead by some measures. They are ahead by other measures." In a "Meet the Press" broadcast, March 6, 1977.

page 41

The 1975 troop comparisons are taken from the Defense Intelligence Agency and the Office of Representative Les Aspin.

page 42

The obsolescence of the Soviet Navy, its limited amphibious capability, and its lack of underway replenishment has been pointed out in Congressional testimony. See the previously cited *Allocation of Resources in the Soviet Union and China*. See these volumes also for testimony of Defense Intelligence Agency officials regarding the inferiority of Soviet tanks, aircraft, and precision-guided artillery and missiles.

page 43

Information and sources on the difficulties of estimating Soviet military expenditures can be found in MILITARY BALANCE, in WORLD MILITARY AND SOCIAL EXPENDITURES, in the annual SIPRI YEARBOOK, in reports of the RAND Corporation (Santa Monica, Calif.), and in Congressional hearings.

Up to the last few years, very little material was available in translation on Soviet military doctrine. The exception to this was the classic volume, *Soviet Military Strategy (Voennaya Strategiya)* by Marshal of the Soviet Union, V.D.

Sokolovski. This has been translated and edited in three editions in the U.S.A. and is highly recommended. For the latest edition, see Harriet Fast Scott, ed., New York: Crane, Russak and Company, 1975. For a recent overview of additional works including a new U.S. Air Force translation, see Paul F. Walker, "Soviet Military Doctrine of the 1970s," *Problems of Communism,* 1978 (forthcoming).

CHAPTER 5

page 51

McGeorge Bundy made a famous statement concerning the minimum amount of nuclear damage required to deter a rational political leader from pursuing a conflict by military means which might lead to nuclear war. Bundy, former Presidential Adviser for National Security Affairs, published an article entitled "To Cap the Volcano" in *Foreign Affairs,* v. 48, n. 1 (October 1969), pp. 1–20. Implying that one nuclear weapon on a major city would be an adequate deterrent, Bundy said (pp. 9–10):

> The neglected truth about the present strategic arms race between the United States and the Soviet Union is that in terms of international political behavior that race has now become almost completely irrelevant . . . In the real world of real political leaders—whether here or in the Soviet Union—a decision that would bring even one hydrogen bomb on one city of one's own country would be recognized in advance as a catastrophic blunder; ten bombs on ten cities would be a disaster beyond history; and a hundred bombs on a hundred cities are unthinkable.

page 53

The appropriation titles are listed in the *Appendix* of the U.S. GOVT BUDGET. The budget programs are described in DEFENSE SECRETARY.

page 56

The detailed estimates of military-related activities of other government departments and agencies are taken from U.S. GOVT BUDGET FY 1978, the main document and the Appendix.

page 58

Line items in the Air Force's Operations and Maintenance, Procurement, and Research Development Test and Evaluation appropriations, each of several hundred million dollars and none explained in detail, are those assumed to be CIA funds.

CHAPTER 6

page 61

The strategic forces of the U.S.A., the U.S.S.R., and the Chinese People's Republic are outlined in JOINT CHIEFS FY 1978 and previous issues. Full and meticulously argued estimates are given also in the SIPRI YEARBOOK. The forces of other nations are found in the MILITARY BALANCE.

page 62

The wide variety of effects of nuclear explosions at every scale are officially and authoritatively described in the Atomic Energy Commission–sponsored volume by S. Glasstone, *Effects of Nuclear Weapons* (Washington, D.C.: GPO, 1977). A briefer and useful technical summary by a physicist of the RAND Corporation, H. Brode, appears in *Annual Review of Nuclear Physics* (Stanford, Calif.: Annual Reviews, 1968), p. 153.

Nuclear tests are counted year by year and summarized in SIPRI YEAR-BOOK, for example, in Appendix 8D in the 1977 issue.

page 69

Long-term effects of fallout and the comparisons between effects of weapons and those of catastrophic accidents and large power plants are discussed in *Nuclear Power: Issues and Choices* (Cambridge, Mass.: Ballinger Publishing Company, 1977), a Report of the Nuclear Energy Policy Study Group.

page 71

True global effects, as distinguished from those to be expected among populations in or near nuclear target regions, are the subject of a study by an expert committee; the volume is called *Long-Term Worldwide Effects of Multiple Nuclear-Weapons Detonations* (Washington, D.C.: National Academy of Sciences, 1975).

page 72

The celebrated "scorpion" metaphor was coined by Robert Oppenheimer in early 1953 in the months after the first test explosion of a large thermonuclear device. He was speaking to the Council of Foreign Relations in New York. For the printed version, see *Foreign Affairs,* v. 31 (July 1953), p. 525.

page 73

A new volume, helpful in teaching strategic nuclear effects and nuclear relations is Paul F. Walker, *Strategic Nuclear Weaponry: The Deadly Calculus*

(Pittsburgh: University of Pittsburgh Press, forthcoming 1978). It includes instructional exercises and a list of pertinent films.

CHAPTER 7

page 74

A very good brief account of the triad and its properties is contained in CBO: *U.S. Strategic Nuclear Forces,* April 1977. A more general account, well written and explicit, is given in chapter 1 of ARMS AND STRATEGY; however, we do not share the author's hopes for a technical solution.

page 75

A *Scientific American* article ("Cruise Missiles," v. 236, n. 2, February 1977) by M.I.T. physicist K. Tsipis gives a good overall story of the new guided cruise missiles. He does not deal, however, with countermeasures, which receive more attention in recent *New York Times* articles and an article by Harvard researcher Robert S. Metzger, "Cruise Missiles: Different Missions, Different Arms Control Impact," *Arms Control Today,* v. 8, n. 1 (January 1978).

page 82

A detailed introductory treatment of anti-submarine warfare for strategic purposes is found in a monograph by SIPRI staff (K. Tsipis and R. Forsberg), *Tactical and Strategic Antisubmarine Warfare* (Cambridge, Mass.: M.I.T. Press, 1974). An authoritative compilation of more specialized papers, both on tactics and on technology, is K. Tsipis, A. Cahn, and B. Feld, eds., *The Future of the Sea-Based Deterrent* (Cambridge, Mass.: M.I.T. Press, 1973), especially chapters 2, 3, and 4. Shadowing and the use of sonar are well-treated in the first source above.

page 83

The submarine bases are discussed in JOINT CHIEFS FY 1978, p. 18. The first base for Trident is now under construction at Bangor, Washington, and a preferred East Coast location has been mentioned at Kings Bay, Georgia. See DEFENSE SECRETARY FY 1978, p. 132. Sub construction, test, training and some central functions are found at Groton, Connecticut in submarine force tradition.

The present low level of station-keeping reliability by the best class of Soviet strategic submarines is reported in JOINT CHIEFS FY 1978, p. 14. This remarkable comment does not seem to have received the attention it deserves.

Estimates from MILITARY BALANCE 1977–78 credit the U.S.S.R. with about 33 nuclear antisubmarine ''hunter-killer'' submarines of the ''November'' and ''Victor'' classes, with antisub torpedoes. These boats are nominally capable of attack on the U.S. Poseidon class of missile subs. But in the official DEFENSE SECRETARY FY 1977 (p. 21) we read explicitly: ''Presently the Victor-class SSN is the most capable Soviet ASW platform. The Victor alone does not pose a threat to our Poseidon force.'' Informed rumor claims that the hull design of these Soviet subs make them noisy under water. Taken with their marginal speed, this implies that they may not be capable of sustained shadowing. They are easy to evade, and they cannot easily pick up the U.S. subs at distance, the faint signals lost in their own noise. Most of the nearly 70 U.S. nuclear antisub submarines now in commission are faster, quieter, and well-armed; they share among them as quarry many fewer U.S.S.R. missile subs at sea.

page 84

The calculus of damage became an advanced topic in the late 1960s. The estimates we use are based on the detailed studies summarized by former Defense Secretary Robert McNamara in his report, DEFENSE SECRETARY FY 1969, pp. 50–58. We applied the method in a simplified way to the U.S.A., by using census figures on population density and the generalized damage areas cited in Chapter 6. The census figures we used come from the distributions given in Chapter 32 of *A Comparative Atlas of the Cities of the U.S.A.* (Minneapolis University of Minnesota Press, 1977). The data come from the 1970 census, but give more detail than the usual list of Standard Metropolitan Areas.

page 87

A useful study of the effects of nuclear attack on points not very near the target is contained in the report by the U.S. Senate, Committee on Foreign Relations, *Analyses of Effects of Limited Nuclear Warfare* (Washington, D.C.: GPO, 1975).

page 89

Current military opinion on the anti-ballistic missile is sketched in JOINT CHIEFS FY 1978, pp. 23–24. Our active defenses are summarized in the same pages, and also in DEFENSE SECRETARY FY 1978, pp. 136–146. On the particle beam, see allegations made in the trade journal, *Aviation Week and Space Technology*.

page 90

Soviet defenses are treated by the JOINT CHIEFS FY 1978, and in MILI-TARY BALANCE 1977–78.

page 91

The citation concerning U.S. early warning systems is from DEFENSE SECRETARY FY 1978, p. 142.

An independent study which comes to the same recommendations we offer for future U.S. strategic offensive forces is that of Peter King, University of Sydney, in *Arms Control and Technological Innovation,* edited by David Carlton and Carlo Schaerf, John Wiley & Sons, New York, 1978. The volume reports other valuable papers from the 1976 International School on Disarmament, Nemi.

CHAPTER 8

page 92

Secretary of State John Foster Dulles invoked our "capacity for massive retaliation . . ." in a speech of January, 1954. See *The Common Defense* cited above for a discussion of the issue.

page 93

The policy of nuclear "options" is outlined with candor in DEFENSE SEC-RETARY FY 1978, pp. 69–79. The language is controlled but the concepts are much less so.

page 94

The decision on the B-1 bomber was taken initially by President Jimmy Carter in late June of 1977. In one or another form, the idea of a low-level penetrating bomber still lives, and the plane may yet reappear in strength, reborn as an FB-111H or even a B-X. See, for example, R. Aldrich, *Nation,* December 3, 1977, p. 592.

page 95

A sketch of the history of MIRV is found in ARMS UNCONTROLLED with a chronology, pp. 134–135. A participant's account can be read in Chapter 9 of H. York's *Race to Oblivion* cited earlier. Another good account is T. Greenwood, *Making the MIRV* (Lexington, Mass.: Ballinger Press, 1975).

page 96

The "Evader" warhead is mentioned in DEFENSE SECRETARY FY 1978, p. 132. See also an article by R. Aldrich, a former project engineer on this scheme, in *Nation*, June 4, 1977.

page 98

A similar analysis of counterforce options is presented by George B. Kistiakowsky, Science Adviser to President Dwight Eisenhower, in the *New York Times Magazine*, November 27, 1977. See the correspondence as well: a rebuttal in brief by Eugene V. Rostow, December 25, 1977, and the rejoinder by Kistiakowsky on January 15, 1978. A fuller statement of the threatening scenario seen by Rostow and his associates is published in *Foreign Affairs*, v. 54, n. 2 (January 1976), pp. 207–232, and in *Foreign Policy* 25 (Winter 1976–77), pp. 195–210, both written by Paul Nitze, a former senior public official and Chairman for Policy Studies of the Committee on the Present Danger. We might and another paper close to our view, by a former CIA deputy director, Herbert Scoville, Jr. See *Foreign Policy* 14 (Spring 1974). pp. 164–177.

CHAPTER 9

page 105

The world manpower, the nature of various weapons and organizations, and other details of this chapter can frequently be traced to sources used for the figures and cited in their captions or notes. We include a few others in these notes where especially helpful.

A similar introductory overview to combat forces is given in chapters 3 and 4 of ARMS AND STRATEGY.

page 109

The present forces of the U.S.A. require about 400,000 new persons entering service each year. The present outlook for recruiting these entrants without conscription is discussed in the *New York Times*, November 15, 1977. A substantial study of the problem, centered on cost-effectiveness, is CBO: *Costs of Defense Manpower*, January 1977.

page 117

A vivid account of the 1973 Yom Kippur war by a senior general of the Israeli Army has been published. The military conclusions from his experience he keeps well-guarded, but the material is striking. See General Chaim Herzog's book cited in Chapter 10 below.

page 118

A future projection which takes little account of the limitations and difficulties of technology is given in a book by Peter Dickson, *The Electronic Battlefield* (Bloomington, Ind: Indiana University Press, 1976).

For recent experiences in air combat, see *Air War in Vietnam,* edited by R. Littauer (Ithaca, N.Y.: Cornell University Press, 1972).

page 119

Air tactics are well summarized in DEFENSE SECRETARY FY 1978, pp. 207–227, and in CBO: *Tactical Air,* January 1977.

page 127

The Air Force language of the "high-low mix" is spelled out in DEFENSE SECRETARY FY 1978, pp. 203–204, and JOINT CHIEFS FY 1978, pp. 80–84.

The appraisal of Soviet aircraft comes from the aircraft specifications (see figure captions for sources) and from JOINT CHIEFS FY 1978, especially pp. 78–80. There is a valuable treatment also in a Brookings Institution study of a few years back, BROOKINGS: *Tactical Air.*

page 129

Basic data on the numbers and types of ships in the navies of the United States, the Soviet Union and other countries are taken from the following sources:

K. Jack Bauer, *Ships of the Navy 1775–1969,* Troy, New York: Rensselaer Polytechnic Institute, 1969.

Alva M. Bowen, "Comparison of U.S. and U.S.S.R. Naval Shipbuilding," Washington, D.C.: Library of Congress, Congressional Research Service, March 5, 1976, processed.

Siegfried Breyer, *Guide to the Soviet Navy,* 1st edition, Annapolis, Maryland: U.S. Naval Institute Press, 1970.

Jean Labayle Couhat, *Combat Fleets of the World 1976/77,* Annapolis, Maryland: U.S. Naval Institute Press, 1976.

Jane's Fighting Ships 1977/78, London: Sampson, Low, Marston, 1977.

Samuel L. Morison and John S. Rowe, *The Ships & Aircraft of the U.S. Fleet,* 10th edition, Annapolis, Maryland: U.S. Naval Institute Press, 1975.

Helpful analyses of the roles and capabilities of various components of the naval forces were found in the following sources:

H.S. Eldredge, "Non-Superpower Sea Denial Capability: The Implications for Superpower Navies Engaged in Presence Operations," paper presented at the Conference on Implications of the Military Build-Up in Non-Industrial States, Boston: Tufts University, Fletcher School of Law and Diplomacy International Security Studies Program, May 4–6, 1976, processed.

Arnold Kuzmack, *Naval Force Levels and Modernization: An Analysis of Shipbuilding Requirements,* Washington, D.C.: Brookings Institution, 1971.

Michael MccGwire and John McDonnell, eds., *Soviet Naval Influence,* New York: Praeger Publishers, 1977.

"U.S. Naval Force Alternatives," Staff Working Paper, Washington, D.C.: U.S. Congress, Congressional Budget Office, March 26, 1976, processed.

CHAPTER 10i

page 143

For the official U.S. Department of Defense description of the European balance, see DEFENSE SECRETARY FY 1978, pp. 93–100 and 108–109. See also DEFENSE SECRETARY FY 1977, pp. 91–93 and 96–100 and preceding annual Defense Department statements.

page 147

NATO and Pact force comparisons are given in MILITARY BALANCE 1976–77, pp. 99 and 104; MILITARY BALANCE 1977–78, pp. 104 and 110; and Robert L. Fischer, "Defending the Central Front: The Balance of Forces," *Adelphi Papers* 127 (London: IISS, 1976), p. 7.

Both France and Denmark have been included in the figures. The numbers for the French forces cover only those stationed in Germany; France is a member of NATO, but its troops are technically not in cooperation with NATO. They are included here because they are under a status agreement with the West German government and would be fielded with NATO during crisis. The Danish forces are oftentimes covered under NATO's Northern Command; they are included here because they face the Polish and East German plains, a potential route of advance of the Pact into the Central Region. Romania and Hungary are excluded because they are not on the Central Front. If sufficient warning were given, both NATO and Pact forces would be expanded; for example, France has an additional 270,000 active Army troops. Whenever such number comparisons are made, one must be careful to note what is included and excluded.

For variations in divisional sizes between countries, see, for example, the Fischer paper cited above, especially p. 8 ff.; also MILITARY BALANCE 1977–78, pp. 92–93 and vii. A good recent study of the NATO-Pact balance is the recent CBO paper, *Assessing the NATO/Warsaw Pact.*

For NATO and Pact equipment estimates, see MILITARY BALANCE 1976–77, p. 101 ff., and MILITARY BALANCE 1977–78, p. 107 ff.

page 150

The NATO and Pact aircraft estimates are taken from MILITARY BALANCE 1976–77, pp. 102–103, and MILITARY BALANCE 1977–78, pp. 107–108.

The combat aircraft characteristics are taken from MILITARY BALANCE 1976–77, p. 74, and MILITARY BALANCE 1977–78, pp. 88–89; also Center for Defense Information, "Military Confrontation in Europe: Will the MBFR Talks Work?" *The Defense Monitor,* v. 4, n. 10 (December 1975), p. 4. Note that an aircraft's maximum range is approximately triple its combat radius depending on the mission. For a fuller description of such aircraft, see JANE'S AIRCRAFT 1976–77.

page 153

The quote regarding Soviet tanks is taken from the Center for Defense Information paper cited above, p. 6.

page 155

The Soviet and NATO tank characteristics are taken from MILITARY BALANCE 1976–77, pp. 89–90; and BROOKINGS: Richard D. Lawrence and Jeffrey Record, *U.S. Force Structure in NATO: An Alternative,* p. 20. The Soviet Union's new T-72 tank, now being introduced into Warsaw Pact forces, is judged inferior to the newest NATO tanks under development, the German Leopard II and the American XM-1; it is judged marginally superior at best to the American M-60 models. See U.S. Congress, Joint Economic Committee, *Allocation of Resources in the Soviet Union and China–1977* (Washington, D.C.: GPO, 1977), p. 74.

page 156

The improbability of a Warsaw Pact attack has been well summarized in MILITARY BALANCE 1977–78, p. 109:

> First, the overall balance is such as to make military aggression appear unattractive. NATO defences are of such a size and quality that any attempt to breach them would require major attack. The consequences for an attacker would be incalculable, and the risks, including that of nuclear escalation, must impose caution. Nor can the theatre be seen in isolation: the central strategic balance and the maritime forces . . . play a vital part in the equation as well.

> . . . while an overall balance can be said to exist today, the Warsaw Pact appears more content with the relationship of forces than is NATO. It is NATO that seeks to achieve equal manpower strengths through equal force reductions while the Pact seeks to maintain the existing correlation.

This and other judgments of an overall NATO-Pact balance and therefore of the improbability of attack directly counters recent warnings of a quick Pact "blitzkrieg" attack; see, for example, an address by Senator Sam Nunn, reprinted in *Survival,* v. 19, n. 1 (January–February 1977), pp. 30–32.

page 158

A recent Congressional study has offered three options for improving the NATO forces, one of which is to "modernize smaller U.S. forces," essentially what we recommend here. The study proposes as one option "deleting the three recently added active divisions from U.S. ground forces." See CBO: *U.S. Air and Ground Conventional Forces for NATO: Overview.*

Information on battlefield engagements from the 1973 Middle East war can be found in Chaim Herzog, *The War of Atonement: October, 1973* (Boston: Little, Brown and Company, 1975); also in Nadav Safran, "Trial by Ordeal: The Yom Kippur War, October 1973," *International Security,* v. 2, n. 2 (Fall 1977), pp. 133–170, and his book, *Israel—The Embattled Ally* (Cambridge, Mass.: Harvard University Press, 1978).

page 162

For more detail on NATO standardization see the report: U.S. House of Representatives, Committee on International Relations, *NATO Standardization: Political, Economic, and Military Issues for Congress* (Washington, D.C.: GPO, March 29, 1977). See also Steven L. Canby, *NATO Military Policy: Obtaining Conventional Comparability with the Warsaw Pact* (Santa Monica, Calif.: RAND Corporation R-1088, June 1973).

CHAPTER 10ii

page 164

The calculations of Soviet military strength appropriate to a discussion of China have been made as follows: (1) the Army was calculated at 450,000 from estimates presented in MILITARY BALANCE, p. 9, of the number of Soviet divisions on the Sino-Soviet border, including the Central Asian, Siberian, Transbaikal, and Far East military districts. These raw division totals have in turn been factored by the degrees of combat readiness, and multiplied by the total number in the Army (p. 8) to arrive at the numerical total. (2) The Air

Force was calculated at 100,000 from estimates (p. 10) of the ratio of the Soviet combat aircraft in military districts outside Eastern Europe multiplied by the total Air Force personnel, but not including the Air Defense Force (PVO-Strany).

No figures for Soviet Navy personnel have been used in figuring these totals, because in the context of a Sino-Soviet confrontation, the Soviet Navy's role is not assumed to be major.

See also text and notes below for a further discussion of this aspect of the Soviet Navy. The respective naval personnel have been included in the armed forces personnel totals for all the other powers listed.

page 165

The U.S.S.R. figures for personnel and total combat aircraft (including the MIG series) are for all locations and include the Air Defense Force as well as the Air Force proper.

The material on the Chinese Navy is based on information in "The PRC Navy—Coastal Defense or Blue Water?" by Commander Bruce Swanson, U.S. Navy, in *Proceedings* (of the United States Naval Institute), v. 102, n. 879 (May 1976), pp. 82–107.

Conclusions regarding the Sino-Soviet border are based partly on MILITARY BALANCE 1977–78, pp. 8–9; taken into account were the numbers of Soviet divisions assigned to each frontier military district, their readiness quotient and the length of the frontier of the particular military district.

page 166

For a detailed listing of the air, land, and naval forces of China and Japan, see MILITARY BALANCE 1977–78, pp. 52–54 (China), 57–58 (Taiwan), and 59–60 (Japan).

page 167

Information on Chinese nuclear developments can be found in MILITARY BALANCE 1977–78, p. 52.

CHAPTER 10iii

page 169

Japanese military forces are given in MILITARY BALANCE 1977–78, pp. 59–60. See also pp. 82–83 for comparisons of percents of GNP devoted to military expenditures.

CHAPTER 10iv

page 175

MILITARY BALANCE 1977–78, p. 83, gives estimates of Iranian arms expenditures for 1974–1977. See also U.S. House of Representatives, Committee on International Relations, *United States Arms Policies in the Persian Gulf and Red Sea Areas: Past, Present, and Future* (Washington, D.C.: GPO, 1977), p. 119, for slightly differing estimates for 1970–1976. See also Figure 47 for U.S. Iranian Foreign Military Sales Agreements and data from *U.S. Arms Policies* cited above, p. 135. For the present composition of Iran's armed forces, see MILITARY BALANCE 1977–78, pp. 35–36.

See MILITARY BALANCE 1977–78 for the armed forces and expenditures of Iran's neighbors, pp. 55–56 (Afghanistan) and 63 (Pakistan).

For a comparison of recent military expenditures of Iraq and Iran, see MILITARY BALANCE 1977–78, p. 83. Iraq's present military forces are included on p. 36.

page 177

For a list of the Iranian arms inventory, see MILITARY BALANCE 1977–78, pp. 35–36, including arms from the U.K., the U.S.A., and the U.S.S.R. See also *U.S. Arms Policies* cited above, pp. 126–127, and for a further list of foreign supplies of Iranian arms, p. 161.

The difficulties of estimating the numbers of U.S. citizens in Iran are considerable. The following estimates are for residents alone and are based on data from *U.S. Arms Policies*, pp. 139–146. (1) U.S. uniformed military technicians and civilian employees of the Department of Defense were about 863 in 1977 and 1,782 in 1978. (2) U.S. citizen employees of U.S. defense contractors were 2,728 in 1975 and are estimated a 4–5,000 for 1977, of whom 3,200 are retired U.S. military employees. (3) U.S. citizens, members of the private sector, including dependents of the above two categories, amounted to about 30,000 in 1977.

It is estimated that the total number of U.S. citizens resident in Iran will rise to 50–60,000 by 1980. The inability of the Iranian armed forces to provide trained personnel to operate sophisticated military equipment, or even recruits for training, is noted throughout the article on Iran, pp. 115–164. In the case of hostilities involving the use of such equipment, a serious problem would arise from the use of U.S. citizens in a combat situation.

CHAPTER 10v

page 179

For an overview of the hundreds of U.S. interventions worldwide, see

BROOKINGS: Barry Blechman and Stephen Kaplan, "Assessing U.S. Uses of Force," 1977.

Two insightful studies discussing problems of the third world and the lack of success of U.S. efforts therein are Douglas S. Blaufarb, *The Counterinsurgency Era: U.S. Doctrine and Performance 1950 to the Present* (New York: Free Press/Macmillan, 1977); and Michael Harrington, *The Vast Majority: A Journey to the World's Poor* (New York: Simon and Schuster, 1977). A third work of possible interest to the reader is Melvin Gurtov, *The United States Against the Third World: Antinationalism and Intervention* (New York: Praeger, 1974).

page 181

Two particularly good and thought-provoking works on U.S. involvement in Vietnam are highly recommended. They vividly point out the falseness in the assumption that firepower alone can win a war in the third world. The first, by a journalist, is Michael Herr, *Dispatches* (New York: Knopf, 1977); the second is by a former Marine officer, Philip Caputo, *A Rumor of War* (New York: Holt, Rinehart and Winston, 1977).

An official Defense Department publication on the Vietnam battlefield experience is John Albright, John A. Cash, and Allan W. Sandstrum, *Seven Firefights in Vietnam* (Washington, D.C.: U.S. Army, 1970).

CHAPTER 11

page 186

The equipment and personnel of U.S. Army divisions of all sorts are detailed in a number of sources, often a little behind the times. These details are in steady change. Important sources are these: for organizations and personnel, see U.S. Army Infantry School, *Infantry Reference Data* (ST-7-157, FY 75), Fort Benning, Geogia; for equipment, except for disposable items (like Redeye missiles!), see U.S. Army Command and General Staff School, *Organizational Data for the Army in the Field* (RB 101-1), June 1974, Fort Leavenworth, Kansas.

The MILITARY BALANCE offers succinct data, but no indication of sources or reliability.

page 188

The current deployment of all services is given in considerable detail in MANPOWER. We have used the text as of FY 1977 (chapter V) with corrections from briefer summaries found in more frequently issued volumes.

page 191

Aircraft disposition is less certain. The squadrons and their deployment—recall how mobile aircraft are—are given in some detail in MANPOWER, Chapter V, and the total inventory in a useful breakdown in Chapter XIII. The number of unit-equipped aircraft per squadron, by type, is not so accessible, especially for U.S. Navy and U.S. Marine Corps squadrons. Various sources, however, allow a good approximation.

page 196

For the sources relevant to analysis of naval force requirements, see the note to page 129 in Chapter 9.

page 203

The airlift and sealift forces are described in DEFENSE SECRETARY FY 1978, pp. 228 ff., in particular the present status of the C-5A heavy-lift jets. The usual reference sources give descriptions of the plane and ship types involved.

page 208

The overseas bases are listed with their functions in a volume of the Department of Defense, *Base Structure Annex*, from the annual MANPOWER report.

CHAPTER 12

page 211

For the official U.S. view of tactical nuclear weapons, see DEFENSE SECRETARY FY 1977, pp. 79–86, and DEFENSE SECRETARY FY 1978, pp. 147–151.

The figures for tactical nuclear weapons deployment are taken from Center for Defense Information, "30,000 U.S. Nuclear Weapons," *The Defense Monitor*, v. 4, n. 2 (February 1975). These numbers are generally accepted as accurate estimates although they have never been officially verified by the Defense Department.

page 213

Data on the range and yield of tactical nuclear weapons may be found in BROOKINGS: Jeffrey Record, *U.S. Nuclear Weapons in Europe: Issues and Alternatives*, 1974, pp. 22 and 24; also MILITARY BALANCE 1977–78, pp. 75 ff. Combat radius is given here for aircraft; if full range were given, the figures would be double or triple. Aircraft ranges also vary with type of mission—load, altitude, environment, loiter time, etc.

There is no official definition for tactical or theater nuclear weapons, as indicated by their omission in the Department of Defense, *Dictionary of Military and Associated Terms* (Washington, D.C.: GPO, 1974). Authors have described them in various ways. For example, John Newhouse in *U.S. Troops in Europe* (Washington, D.C.: BROOKINGS, 1971) states (p. 45) that tactical nuclear weapons are "designed to influence the land or air defense battle directly, rather than indirectly (for example, by interdicting an enemy's lines of communication or by carrying a threat to his homeland)." Similarly, Morton H. Halperin describes them as "nuclear weapons designed to support land forces." See his *Defense Strategies for the Seventies* (Boston: Little, Brown and Company, 1971), p. 5.

page 214

For Soviet military doctrine regarding nuclear weapons, see V.D. Sokolovski, *Soviet Military Strategy,* translated by H.F. Scott (New York: Crane, Russak and Company, 1975). See also the "Soviet Military Thought" series, translated by the U.S. Air Force and available from the U.S. GPO; also Joseph D. Douglass, Jr., *The Soviet Theater Nuclear Offensive* (Washington, D.C.: GPO, 1976).

The effects of nuclear weapons in limited exchanges are described in U.S. Senate, Committee on Foreign Relations, *Analyses of Effects of Limited Nuclear Warfare* (Washington, D.C.: GPO, 1975). For information on Carte Blanche, see Helmut Schmidt, *Defense or Retaliation* (Edinburgh: Oliver and Boyd, 1962).

page 215

For a recent critique of U.S. tactical nuclear weapons policy in Europe, see Jeffrey Record, "Theatre Nuclear Weapons: Begging the Soviet Union to Preempt," *Survival,* v. 19, n. 5 (Septmeber–October 1977), pp. 208–211.

page 218

Additional, concise information on the neutron bomb or enhanced radiation weaponry can be found in Harold M. Agnew, "A Primer on Enhanced Radiation Weapons" and Bert Sorensen, "No Neutron Bombs for Us, Please," in *The Bulletin of the Atomic Scientists,* v. 33, n. 10 (December 1977), pp. 6–8. The Soviets are beginning to deploy a mobile, intermediate-range ballistic missile, the SS-20, in the Western U.S.S.R. targeted on Europe; although this does not appear to destabilize the European balance, it does appear as an unnecessary and escalating step.

CHAPTER 13

page 223

This chapter deals with the most complex and least open portion of the forces. We have used various schemes to rough out the structure. The public interest and the Congressional investigations of the last few years have given us enough insight to go ahead. Recent annual reports of the Department of Defense have included quite good reviews of structure and function. In particular, the complex modern systems of command, communication, and control are diagrammed and described very well in DEFENSE SECRETARY FY 1977, Chapter V. Our simplified diagram comes from this source.

page 225

Chapter 6 of ARMS UNCONTROLLED is a handy description of underwater surveillance systems, both fixed and mobile.

CHAPTER 14

page 232

For the annual official Defense Department explanation of R&D, the reader should obtain the lengthy statement prepared by the Director of Defense Research and Engineering, PROGRAM RESEARCH. This report details the various R&D budget categories, for example, exploratory development and advanced development.

page 237

The division of R&D funds between universities, Defense laboratories, and industry is given in Appendix A-1 of PROGRAM RESEARCH FY 1978.

page 238

The funding of each individual project was obtained from Defense Department documents, namely PROGRAM ACQUISITION FY 1978 and DEFENSE SECRETARY FY 1978 as well as the budget amendments for FY 1978 (*Amended Budget RDT&E Program—R-1*); the article, "Major Weapon System Spending Detailed," from *Aviation Week and Space Technology,* v. 106, n. 4 (January 24, 1977), pp. 22–23, was also utilized.

The proposed budget was derived as described in Chapter 19. Each item was separately assessed in accordance with our recommended policies outlined in this and other chapters. Approximately 75 percent of the spending could be accounted for in individual procurement requests. The remainder in each category was treated in the same way as the overall average of the identified items. For example, if a 33 percent cut was indicated among the identified items, then a 33 percent cut was also applied to the unidentified remainder.

CHAPTER 15

page 240

This chapter relies heavily on the study of the nature and quality of the Reserves made some time ago and published as BROOKINGS: *Reserve Forces*.

page 243

Up-to-date numbers of persons and costs of the various organizations have been compiled from MANPOWER FY 1977, especially its chapters on the individual services. These in turn were modified by the later data of DEFENSE SECRETARY FY 1978 and the first budget of President Jimmy Carter.

CHAPTER 16

page 249

For the official discussion of support activities, see the section in the annual report of the Secretary of Defense concerning "mission support." This discusses such areas as "central supply and maintenance," "defense logistics," "administration and associated activities," and "training, medical, and other general personnel activities."

Data on support services and personnel are taken from DEFENSE SECRETARY FY 1978.

For a thoughtful and critical examination of support services in Defense, see BROOKINGS: *Support Costs*. The author points out that support programs are "expensive" but are "difficult to examine . . . because bureaucratic, political, economic and timing constraints tend to silence debate over support issues and spending levels." Nevertheless, the author proposes several "alternative support programs" for the military.

CHAPTER 17

page 254

The most recent Defense statements on arms trade are in DEFENSE SECRETARY FY 1977, pp. 185–194 and DEFENSE SECRETARY FY 1978, pp. 238–246.

The figures for "Support of Other Nations" were taken from DEFENSE SECRETARY FY 1978, p. A–1.

page 255

For a graphic illustration of sales and grants since 1950, see DEFENSE SECRETARY FY 1978, p. 244.

page 256

The recipients of American weaponry may be found in ACDA: *World Military Expenditures and Arms Transfers 1966–1975,* 1976, pp. 77–80. This government volume includes a relatively good statistical compilation of worldwide arms transfers by both suppliers and recipients; see also the two earlier volumes for 1963–1973 and 1965–1974, which succeeded the classified volume, *World Arms Trade,* and the unclassified *World Military Expenditures.*

For the spread of sophisticated weaponry worldwide, see SIPRI YEARBOOK. See also the recent illustrated volume by Tom Gervasi, *Arsenal of Democracy: American Weapons Available for Export* (New York: Grove Press, 1978).

page 260

It should be noted that costs of weapon systems vary greatly depending on the contractual arrangements and what is included in the price, for example, spare parts, service, or instruction. The prices cited here were gathered from a variety of public sources.

page 261

A recent comprehensive report on the U.S.-Iran arms trade is U.S. Senate, Committee on Foreign Relations, *U.S. Military Sales to Iran* (Washington, D.C.: GPO, 1976). For an early report on worldwide U.S. arms transfers, see U.S. House of Representatives, Committee on Foreign Affairs, *The International Transfer of Conventional Arms* (a report to Congress from the U.S. Arms Control and Disarmament Agency) (Washington, D.C.: GPO, 1974).

page 262

Figures on deliveries to geographic regions are given in ACDA: *World Military Expenditures and Arms Transfers 1966–1975*, especially pp. 1–3 which give military spending by country and by region.

page 266

A recent United Nations Association report similarly recommended, among other things, that "the U.S. should exercise greater restraint in its arms export policies" and "should put its full weight behind regional initiatives towards controlling the arms trade.'' See United Nations Association of the U.S.A., ''Revised Report of the Arms Trade Sub-panel,'' New York, March 5, 1976, p. 7.

The quotes on arms trade are from DEFENSE SECRETARY FY 1977, p. 189.

page 268

The Carter administration recognized the gravity of the problem in mid-1977 American trade in weaponry. As Carter spoke in his campaign, the U.S.A. cannot be ''both the world's leading champion of peace and the world's leading supplier of weapons of war.'' Quoted in Leslie H. Gelb, ''Arms Sales,'' *Foreign Policy* n. 25 (Winter 1976–77), p. 3.

CHAPTER 18

page 269

These points are emphasized in the last Defense report; see the section on ''manpower'' in DEFENSE SECRETARY FY 1978, pp. 284–302. See also MANPOWER FY 1978.

page 270

See the third Quadrennial Review of Military Compensation, initiated by the Secretary of Defense upon the direction of the President in January, 1975; reported in ''A Comparison of Proposed Changes in Military Retirement, Compensation,'' *Army,* v. 27, n. 2 (February 1977), pp. 29–32. See also CBO: *Military Retirement System,* 1978. This report recommended an overhaul of the pension system.

page 271

The Carter administration recognized the gravity of the problem in mid-1977 when a new Presidentially-appointed Commission on Military Compensation was constituted to examine retirement and pay problems. In testimony before the Commission, many top military officers have stated that it is time for ''a fairer and more flexible retirement policy.'' See the *New York Times,* January 19, 1978, p. 33. The commission's report was made in April, 1978.

page 272

The average ages of retired military were reported from the Office of Congressman Les Aspin (D-Wisconsin) in 1977.

page 274

For a recent study of the compensation issue, see, for example, BROOK-INGS: Martin Binkin, *The Military Pay Muddle,* 1975.

The figures on per capita costs are from the Departments of Defense and Commerce, as reported in CBO: *Defense Manpower: Compensation Issues for Fiscal Year 1977,* 1976, p. 7.

A recent Congressional report indicates that military benefits need examination. See CRS: "What's Happened to Military Pay and Benefits Through the Past Decade?" 1977.

page 276

Personnel costs for 1978 are taken from DEFENSE SECRETARY FY 1978, p. 287.

CHAPTER 19

page 279

For the 1978 budget, see DEFENSE SECRETARY FY 1978, Appendix C, especially pp. C–11 to C–14.

Extensive information on the annual procurement and research and development programs can be found in the annual hearings before Congress. See also two annual Defense Department reports: PROGRAM RESEARCH and PROGRAM ACQUISITION. The defense budget estimate for fiscal year 1978 used throughout this book, $120.4 billion, represents obligational authority proposed by the outgoing Ford Administration and then amended by the new Carter Administration. The budget eventually passed by Congress for FY 1978 is reported as $116.8 billion. Most of the difference between the two is accounted for by the cancellation of the B-1 bomber (a saving of $1.6 billion) and the transfer of military assistance funds ($1.3 billion) from the defense accounts to the Budget of the State Department. As of August 1978, it appears that the comparable budget for FY 1979 will be sharply up, by about $10 billion.

page 287

The following table presents a sampling of the myriad decisions made by the authors in estimating R & D and Procurement budgets.

Table A. Selected Budget Decisions in Procurement

Air Force	*Army*	*Navy*
	Items Cut 100%	
B-1 Bomber	Mechanized Infantry	Trident Submarine
E-3A AWACS Aircraft	Combat Vehicle	Nuclear Aircraft
Minuteman ICBM	(MICV)	Carrier
Air-Launched Cruise	UTTAS Helicopter	Aegis Destroyer
Missile	Self-Propelled	Attack Submarine
	Howitzer	F-18 Tactical Aircraft
	Lance Missile	
	Items Partially Cut	
F-15 Tactical Aircraft	Common Ground	Missile Modification
Aircraft Support	Support	Aircraft Support
Equipment	Spares and Repair	Aerial Targets
	Parts	TOW Missile
	Ammunition	(Marines)
	Items Retained at Present or Higher Levels	
F-16 Tactical Aircraft	M-60 Tank	Poseidon Tests and
A-10 Tactical Aircraft	M-48 Tank	Modifications
Sparrow Missile	AH-15 Cobra/TOW	Harpoon Missile
Sidewinder Missile	Helicopter	MK-46 Torpedo
Maverick Missile	Stinger Missile	Shrike Missile

Table B. Selected Budget Decisions in R&D

Air Force	*Army*	*Navy*
	Items Cut 100%	
B-52 Squadrons	Tank Gun	Trident Missile
Short-Range Attack	UTTAS Transport	A-6 Aircraft
Missile	Helicopter	Cruise Missile
Advanced ICBM	CH-47 Helicopter	AEGIS Weapon
B-1 Bomber	Pershing Missile	System
KC-135 Squadrons	Mechanized Infantry	Marine
	Combat Vehicle	Telecommunications
		Satellite
		Communications

(continued)

Air Force	Army	Navy
	Items Partially Cut	
Drone Support	Tactical	Undersea Surveillance
Defense Satellite	Communications	Tactical Information
Communications	Advanced Air Defense	Fleet
		Telecommunications

Items Retained at Present or Higher Levels

F-15 Tactical Aircraft	Anti-Tank Assault	SSBN Security
A-10 Tactical Aircraft	Chapparal Missile	Submarine Silencing
F-16 Tactical Aircraft	Cobra TOW Missile	Submarine Sonor
SLBM Radar Warning	Hellfire Missile	Airborne ASW
	Patriot SAM D Missile	

page 288

The budget matrix methodology was developed by Randall Forsberg on the basis of detailed estimates published in SECRETARY DEFENSE, U.S. GOVT BUDGET, *Air Force Magazine* annual Almanac Issue and U.S. Navy force and budget statistics. The full detail of the matrix and explanation of the sources and methods are set out in the forthcoming booklet *The Structure of U.S. Military Spending*, R. Forsberg, Institute for Policy Studies.

An example of the budget methodology is given below. It presents four squares of the large matrix developed by aligning appropriation titles with military programs. This allowed hundreds of figures, broken out from the larger programs, to be determined by estimated percentages. We make no claims of exactness, but this procedure allowed more accuracy than if one worked only with gross amounts.

Table C. Sample of FY 1978 Defense Budget Breakdown
(million current $)

Appropriation Title:	*Military Personnel*	*Operations & Maintenance*
Military Program:		
Strategic Forces	2 (Army)	0 (Army)
	236 (Navy)	855 (Navy)
	1538 (Air Force)	1659 (Air Force)
	1776 total	2514 total

General Purpose		
Forces	5601 (Army)	2758 (Army)
	3299 (Navy)	5073 (Navy)
	1253 (Marines)	349 (Marines)
	1857 (Air Force)	1574 (Air Force)
	12,010 total	9757 total

For an earlier extensive study on defense support costs, see BROOKINGS: *Support Costs,* 1972. See also the annual analysis by BROOKINGS: *Setting National Priorities,* which examines the complete national budget.

CHAPTER 20

page 292

In FY 1971, an estimated 6.1 million Americans were in defense-related employment. See ACDA: *The Economic Impact of Reductions in Defense Spending,* 1972, p. 3. There has been real growth in the military budget since, and for purposes of this discussion a conservative (and in this context slightly higher) estimate has been used.

For localized impacts of defense cutbacks, the reader is referred to a number of case studies done in the 1960s and 1970s by the U.S. Arms Control and Disarmament Agency. The Carter Administration is now showing increased interest in further serious study of this important phenomenon.

The $70 billion proposed budget figure utilized in this chapter is rounded for purposes of the economic discussion.

page 296

Details of the B-1 bomber decision and its developer, Rockwell International, can be found in articles in the *New York Times* throughout June and July, 1977. Robert Anderson, President of Rockwell, commented contrary to some press reports that the bomber cancellation would have only "a minimal effect on earnings this year. However, the company will lose the long-term potential which production had offered us." See the *New York Times,* July 1, 1977, p. D1. The Rockwell case is a good example of how well a diversified company can absorb defense cutbacks; the B-1 project was less than 5 percent of Rockwell sales. About 22 percent of its sales are for automobile products, 27 percent for aerospace products, 20 percent in electronics, and the remainder in a variety of different fields.

The Bureau of Labor Statistics study was cited in *Priorities* (Pasadena, California: American Friends Service Committee, June, 1976), p. 3. This material was also used as the basis for Figure 77.

Another study which amplifies the case argued here is the excellent published analysis of Dr. Roger H. Bezdek in *Journal of Regional Science,* v. 15, n. 2 (1975), pp. 183–198. Bezdek tabulates and analyzes the effect of increased and decreased military expenditures on industries, regions, and selected occupations. Figure 78 is based on Bezdek's Table 4, p. 193.

EPILOGUE

page 311

The membership of House and Senate committees, as well as other valuable and related information, can be found in the annual *Official Congressional Directory* (Washington, D.C.: GPO) or in the widely available annual almanacs. The eight committees referred to here are for the House: Appropriations, Armed Services, Budget, and International Relations; and, for the Senate: Appropriations, Armed Services, Budget, and Foreign Relations.

The Congressional schedule for annual budget consideration is based on information in *Washington Newsletter* (Washington, D.C.: Friends Committee on National Legislation, January, 1976), n. 377, p. 3; see also "Congressional Budget Process," *Press Release* of the Senate Budget Committee, December, 1976, n. 76–54, Washington, D.C., pp. 1–4.

page 314

"Let us raise a standard to which the wise and honest can repair": George Washington, 1787, at the Constitutional Convention.

INDEX